MW01383888

The Scientist or Engineer as an Expert Witness

CHEMICAL INDUSTRIES

A Series of Reference Books and Textbooks

1. *Fluid Catalytic Cracking with Zeolite Catalysts,* Paul B. Venuto and E. Thomas Habib, Jr.
2. *Ethylene: Keystone to the Petrochemical Industry,* Ludwig Kniel, Olaf Winter, and Karl Stork
3. *The Chemistry and Technology of Petroleum,* James G. Speight
4. *The Desulfurization of Heavy Oils and Residua,* James G. Speight
5. *Catalysis of Organic Reactions,* edited by William R. Moser
6. *Acetylene-Based Chemicals from Coal and Other Natural Resources,* Robert J. Tedeschi
7. *Chemically Resistant Masonry,* Walter Lee Sheppard, Jr.
8. *Compressors and Expanders: Selection and Application for the Process Industry,* Heinz P. Bloch, Joseph A. Cameron, Frank M. Danowski, Jr., Ralph James, Jr., Judson S. Swearingen, and Marilyn E. Weightman
9. *Metering Pumps: Selection and Application,* James P. Poynton
10. *Hydrocarbons from Methanol,* Clarence D. Chang
11. *Form Flotation: Theory and Applications,* Ann N. Clarke and David J. Wilson
12. *The Chemistry and Technology of Coal,* James G. Speight
13. *Pneumatic and Hydraulic Conveying of Solids,* O. A. Williams

14. *Catalyst Manufacture: Laboratory and Commercial Preparations,* Alvin B. Stiles
15. *Characterization of Heterogeneous Catalysts,* edited by Francis Delannay
16. *BASIC Programs for Chemical Engineering Design,* James H. Weber
17. *Catalyst Poisoning,* L. Louis Hegedus and Robert W. McCabe
18. *Catalysis of Organic Reactions,* edited by John R. Kosak
19. *Adsorption Technology: A Step-by-Step Approach to Process Evaluation and Application,* edited by Frank L. Slejko
20. *Deactivation and Poisoning of Catalysts,* edited by Jacques Oudar and Henry Wise
21. *Catalysis and Surface Science: Developments in Chemicals from Methanol, Hydrotreating of Hydrocarbons, Catalyst Preparation, Monomers and Polymers, Photocatalysis and Photovoltaics,* edited by Heinz Heinemann and Gabor A. Somorjai
22. *Catalysis of Organic Reactions,* edited by Robert L. Augustine
23. *Modern Control Techniques for the Processing Industries,* T. H. Tsai, J. W. Lane, and C. S. Lin
24. *Temperature-Programmed Reduction for Solid Materials Characterization,* Alan Jones and Brian McNichol
25. *Catalytic Cracking: Catalysts, Chemistry, and Kinetics,* Bohdan W. Wojciechowski and Avelino Corma
26. *Chemical Reaction and Reactor Engineering,* edited by J. J. Carberry and A. Varma
27. *Filtration: Principles and Practices: Second Edition,* edited by Michael J. Matteson and Clyde Orr
28. *Corrosion Mechanisms,* edited by Florian Mansfeld
29. *Catalysis and Surface Properties of Liquid Metals and Alloys,* Yoshisada Ogino
30. *Catalyst Deactivation,* edited by Eugene E. Petersen and Alexis T. Bell
31. *Hydrogen Effects in Catalysis: Fundamentals and Practical Applications,* edited by Zoltán Paál and P. G. Menon
32. *Flow Management for Engineers and Scientists,* Nicholas P. Cheremisinoff and Paul N. Cheremisinoff
33. *Catalysis of Organic Reactions,* edited by Paul N. Rylander, Harold Greenfield, and Robert L. Augustine
34. *Powder and Bulk Solids Handling Processes: Instrumentation and Control,* Koichi Iinoya, Hiroaki Masuda, and Kinnosuke Watanabe
35. *Reverse Osmosis Technology: Applications for High-Purity-Water Production,* edited by Bipin S. Parekh
36. *Shape Selective Catalysis in Industrial Applications,* N. Y. Chen, William E. Garwood, and Frank G. Dwyer

37. *Alpha Olefins Applications Handbook,* edited by George R. Lappin and Joseph L. Sauer
38. *Process Modeling and Control in Chemical Industries,* edited by Kaddour Najim
39. *Clathrate Hydrates of Natural Gases,* E. Dendy Sloan, Jr.
40. *Catalysis of Organic Reactions,* edited by Dale W. Blackburn
41. *Fuel Science and Technology Handbook,* edited by James G. Speight
42. *Octane-Enhancing Zeolitic FCC Catalysts,* Julius Scherzer
43. Oxygen in Catalysis, Adam Bielanski and Jerzy Haber
44. *The Chemistry and Technology of Petroleum: Second Edition, Revised and Expanded,* James G. Speight
45. *Industrial Drying Equipment: Selection and Application,* C. M. van't Land
46. *Novel Production Methods for Ethylene, Light Hydrocarbons, and Aromatics*, edited by Lyle F. Albright, Billy L. Crynes, and Siegfried Nowak
47. *Catalysis of Organic Reactions*, edited by William E. Pascoe
48. *Synthetic Lubricants and High-Performance Functional Fluids,* edited by Ronald L. Shubkin
49. *Acetic Acid and Its Derivatives*, edited by Victor H. Agreda and Joseph R. Zoeller
50. *Properties and Applications of Perovskite-Type Oxides,* edited by L. G. Tejuca and J. L. G. Fierro
51. *Computer-Aided Design of Catalysts,* edited by E. Robert Becker and Carmo J. Pereira
52. *Models for Thermodynamic and Phase Equilibria Calculations,* edited by Stanley I. Sandler
53. *Catalysis of Organic Reactions*, edited by John R. Kosak and Thomas A. Johnson
54. *Composition and Analysis of Heavy Petroleum Fractions,* Klaus H. Altgelt and Mieczyslaw M. Boduszynski
55. *NMR Techniques in Catalysis,* edited by Alexis T. Bell and Alexander Pines
56. *Upgrading Petroleum Residues and Heavy Oils,* Murray R. Gray
57. *Methanol Production and Use*, edited by Wu-Hsun Cheng and Harold H. Kung
58. *Catalytic Hydroprocessing of Petroleum and Distillates,* edited by Michael C. Oballah and Stuart S. Shih
59. *The Chemistry and Technology of Coal: Second Edition, Revised and Expanded,* James G. Speight
60. *Lubricant Base Oil and Wax Processing*, Avilino Sequeira, Jr.
61. *Catalytic Naphtha Reforming: Science and Technology,* edited by George J. Antos, Abdullah M. Aitani, and José M. Parera

62. *Catalysis of Organic Reactions*, edited by Mike G. Scaros and Michael L. Prunier
63. *Catalyst Manufacture,* Alvin B. Stiles and Theodore A. Koch
64. *Handbook of Grignard Reagents,* edited by Gary S. Silverman and Philip E. Rakita
65. *Shape Selective Catalysis in Industrial Applications: Second Edition, Revised and Expanded,* N. Y. Chen, William E. Garwood, and Francis G. Dwyer
66. *Hydrocracking Science and Technology,* Julius Scherzer and A. J. Gruia
67. *Hydrotreating Technology for Pollution Control: Catalysts, Catalysis, and Processes,* edited by Mario L. Occelli and Russell Chianelli
68. *Catalysis of Organic Reactions,* edited by Russell E. Malz, Jr.
69. *Synthesis of Porous Materials: Zeolites, Clays, and Nanostructures,* edited by Mario L. Occelli and Henri Kessler
70. *Methane and Its Derivatives,* Sunggyu Lee
71. *Structured Catalysts and Reactors,* edited by Andrzej Cybulski and Jacob A. Moulijn
72. *Industrial Gases in Petrochemical Processing,* Harold Gunardson
73. *Clathrate Hydrates of Natural Gases: Second Edition, Revised and Expanded,* E. Dendy Sloan, Jr.
74. *Fluid Cracking Catalysts,* edited by Mario L. Occelli and Paul O'Connor
75. *Catalysis of Organic Reactions,* edited by Frank E. Herkes
76. *The Chemistry and Technology of Petroleum: Third Edition, Revised and Expanded,* James G. Speight
77. *Synthetic Lubricants and High-Performance Functional Fluids: Second Edition, Revised and Expanded,* Leslie R. Rudnick and Ronald L. Shubkin
78. *The Desulfurization of Heavy Oils and Residua, Second Edition, Revised and Expanded,* James G. Speight
79. *Reaction Kinetics and Reactor Design: Second Edition, Revised and Expanded,* John B. Butt
80. *Regulatory Chemicals Handbook,* Jennifer M. Spero, Bella Devito, and Louis Theodore
81. *Applied Parameter Estimation for Chemical Engineers,* Peter Englezos and Nicolas Kalogerakis
82. *Catalysis of Organic Reactions,* edited by Michael E. Ford
83. *The Chemical Process Industries Infrastructure: Function and Economics,* James R. Couper, O. Thomas Beasley, and W. Roy Penney
84. *Transport Phenomena Fundamentals,* Joel L. Plawsky

85. *Petroleum Refining Processes*, James G. Speight and Baki Özüm
86. *Health, Safety, and Accident Management in the Chemical Process Industries*, Ann Marie Flynn and Louis Theodore
87. *Plantwide Dynamic Simulators in Chemical Processing and Control*, William L. Luyben
88. *Chemical Reactor Design*, Peter Harriott
89. *Catalysis of Organic Reactions*, edited by Dennis G. Morrell
90. *Lubricant Additives: Chemistry and Applications*, edited by Leslie R. Rudnick
91. *Handbook of Fluidization and Fluid-Particle Systems*, edited by Wen-Ching Yang
92. *Conservation Equations and Modeling of Chemical and Biochemical Processes*, Said S. E. H. Elnashaie and Parag Garhyan
93. *Batch Fermentation: Modeling, Monitoring, and Control*, Ali Çinar, Gülnur Birol, Satish J. Parulekar, and Cenk Ündey
94. *Industrial Solvents Handbook, Second Edition*, Nicholas P. Cheremisinoff
95. *Petroleum and Gas Field Processing*, H. K. Abdel-Aal, Mohamed Aggour, and M. Fahim
96. *Chemical Process Engineering: Design and Economics*, Harry Silla
97. *Process Engineering Economics*, James R. Couper
98. *Re-Engineering the Chemical Processing Plant: Process Intensification*, edited by Andrzej Stankiewicz and Jacob A. Moulijn
99. *Thermodynamic Cycles: Computer-Aided Design and Optimization*, Chih Wu
100. *Catalytic Naphtha Reforming: Second Edition, Revised and Expanded*, edited by George T. Antos and Abdullah M. Aitani
101. *Handbook of MTBE and Other Gasoline Oxygenates*, edited by S. Halim Hamid and Mohammad Ashraf Ali
102. *Industrial Chemical Cresols and Downstream Derivatives*, Asim Kumar Mukhopadhyay
103. *Polymer Processing Instabilities: Control and Understanding*, edited by Savvas Hatzikiriakos and Kalman B. Migler
104. *Catalysis of Organic Reactions*, John Sowa
105. *Gasification Technologies: A Primer for Engineers and Scientists*, edited by John Rezaiyan and Nicholas P. Cheremisinoff
106. *Batch Processes*, edited by Ekaterini Korovessi and Andreas A. Linninger
107. *Introduction to Process Control*, Jose A. Romagnoli and Ahmet Palazoglu

108. *Metal Oxides: Chemistry and Applications*, edited by J. L. G. Fierro
109. *Molecular Modeling in Heavy Hydrocarbon Conversions*, Michael T. Klein, Ralph J. Bertolacini, Linda J. Broadbelt, Ankush Kumar and Gang Hou
110. *Structured Catalysts and Reactors, Second Edition*, edited by Andrzej Cybulski and Jacob A. Moulijn
111. *Synthetics, Mineral Oils, and Bio-Based Lubricants: Chemistry and Technology*, edited by Leslie R. Rudnick
112. *Alcoholic Fuels*, edited by Shelley Minteer
113. *Bubbles, Drops, and Particles in Non-Newtonian Fluids, Second Edition*, R. P. Chhabra
114. *The Chemistry and Technology of Petroleum, Fourth Edition*, James G. Speight
115. *Catalysis of Organic Reactions*, edited by Stephen R. Schmidt
116. *Process Chemistry of Lubricant Base Stocks*, Thomas R. Lynch
117. *Hydroprocessing of Heavy Oils and Residua*, edited by James G. Speight and Jorge Ancheyta
118. *Chemical Process Performance Evaluation*, Ali Cinar, Ahmet Palazoglu, and Ferhan Kayihan
119. *Clathrate Hydrates of Natural Gases, Third Edition*, E. Dendy Sloan and Carolyn Koh
120. *Interfacial Properties of Petroleum Products*, Lilianna Z. Pillon
121. *Process Chemistry of Petroleum Macromolecules*, Irwin A. Wiehe
122. *The Scientist or Engineer as an Expert Witness*, James G. Speight

The Scientist or Engineer as an Expert Witness

James G. Speight
University of Trinidad and Tobago
O'Meara Campus, Trinidad

CRC Press is an imprint of the
Taylor & Francis Group, an **informa** business

CRC Press
Taylor & Francis Group
6000 Broken Sound Parkway NW, Suite 300
Boca Raton, FL 33487-2742

Printed in the United States of America on acid-free paper
10 9 8 7 6 5 4 3 2 1

International Standard Book Number-13: 978-1-4200-5258-9 (Hardcover)

Library of Congress Cataloging-in-Publication Data

Speight, J. G.
The scientist or engineer as expert witness / James G. Speight.
p. cm. -- (Chemical industries ; 122)
Includes bibliographical references and index.
ISBN 978-1-4200-5258-9 (alk. paper)
1. Evidence, Expert--United States. 2. Forensic sciences--United States. 3. Forensic engineering--United States. I. Title.

KF8968.66.S64 2009
347.73'67--dc22 2008029653

Visit the Taylor & Francis Web site at
http://www.taylorandfrancis.com

and the CRC Press Web site at
http://www.crcpress.com

Contents

Preface

There are only a few books that relate to the expert witness. None appears to be aimed specifically at the scientist and engineer. Many of the books are written for the legal profession and cite many court cases, rules, and laws.

As the number of litigation cases increases and more cases than ever before are based on technical subjects, a book written specifically for scientists and engineers is necessary. The scientist and engineer present the salient technical facts to the court in areas varying from toxicity, other adverse effects of a chemical or physical agent, and immediate injury due to explosions and fire to chemicals splattered on the face with damage to skin or eyes. In addition, harm caused by exposure to a chemical substance may not manifest itself for a number of years, and individual plaintiffs may not display specific symptoms at the time of a trial. The claims for damage can be for a variety of effects, and all members of the plaintiffs' case may not claim the same injury. Also to be considered is the individual who is sensitive to a specific compound. This person may seek monetary damages against the manufacturer of that agent.

All of these effects may fall into the area of science and engineering in which the science or engineer must present the truth clearly and coherently to the court. The scientist or engineer has the task of explaining to a judge and jury exactly what caused the incident and the effects of the chemical under question. The mathematical equations propagated by many engineers may not suffice. The data must be presented in an understandable and unbiased manner to enable nonscientists (the judge and jury) to make a decision.

The scientist and engineer are also needed in regulatory rule making. Experts are often called before legislative committees at local, state, or national levels to present evidence that will lead to new regulations and laws. There are also cases where scientists and engineers are needed at trials for whistle-blowers as well as for cases of scientific misconduct. Some states mandate that an expert be used in certain cases, especially if another expert is being sued for professional negligence or scientific misconduct.

Not many scientists and engineers have the experience to testify effectively. Not all attorneys know how to deal with competent scientists and engineers. This is a learning experience for all involved.

This book is aimed at scientists and engineers who intend to be useful as experts for the plaintiff or for the defense. Someone—the attorney, the paralegal, the scientist, or the engineer—must take on the responsibility for presentation of the data to the court. Scientists and engineers must attempt to be the most effective witnesses possible. In the courtroom, they are mainly teachers and explainers of complicated phenomena.

In addition, both plaintiff and defense lawyers and their paralegal assistants can profit from this book, which is written from the viewpoint of an experienced scientist.

James G. Speight

The Author

James G. Speight, PhD, DSc, has more than 40 years' experience in areas associated with the properties and processing of conventional and synthetic fuels. He has participated in, as well as led, significant research in defining the processing chemistry of petroleum, natural gas, heavy oil, tar sand bitumen, coal, oil shale, and their respective products. He has provided more than 400 publications, reports, and presentations detailing these research activities and has taught more than 70 related courses. He has also been called as an expert witness on areas related to business and fossil fuel technology.

Dr. Speight is currently editor of the journals *Petroleum Science and Technology* (formerly *Fuel Science and Technology International*); *Energy Sources Part A: Recovery, Utilization, and Environmental Effects*; and *Energy Sources Part B: Economics, Planning, and Policy*. He is recognized as a world leader in the areas of fuel characterization and development. Dr. Speight is also visiting professor at the University of Trinidad and Tobago and adjunct professor of Chemical and Fuels Engineering at the University of Utah.

1 Being an Expert Witness

1.1 INTRODUCTION

Over the past three decades, the courts have become inundated with disputes that involve scientific and engineering principles, and there is a need for qualified scientists and engineers to present testimony to the court for the plaintiff or for the defendant. Furthermore, these disputes involving the need for technical experts promise to continue for the foreseeable future.

This scientific and engineering aspect of legal proceedings is not new; it has been on court dockets for decades but for the most part came to the foreground in the 1980s as a result of various environmental issues. The technical aspects of environmental science and engineering were no longer the domain of the uninitiated layperson but became the domain of the trained scientist and engineer. This is not an attempt to deny the occurrence of the technical needs of other cases but to acknowledge the lack of expert technical support in many court cases.

The need for expert testimony is acute now that court cases and disputes have expanded and vary from interpretation of language and definitions to patent infringement issues. The technical evidence can help determine who is at fault. Furthermore, unlike many of the rules of evidence adopted by many states, Rule 803 of the Federal Rules of Evidence provides that an expert may base his opinion on facts and documents not in evidence, as long as those facts and documents are reasonably relied upon by experts in his field. In this respect, the scientist or engineer is called to present the truth, as he or she sees it, to the court and not to make any judgment as an advocate for either side.

On the other hand, the client who insists, for some reason, that the expert be an advocate is moving beyond the realms of the definition of an expert witness by asking the expert to take a stance on the issues rather than presenting the unbiased facts to the court. The difference between being an expert witness and an advocate can be likened to the difference between honesty and perjury.

This phenomenon has created a new role for scientists and engineers because they now venture into the unfamiliar ground of the courtroom but still have to present the necessary information to the court in a manner that is both logical and understandable. The best witness on earth will fail if he or she presents evidence that is not understood by the court.

Most attorneys are comfortable using scientists and engineers as expert witnesses because these professions are well known and characterized. Normally, members of a jury are acquainted with one who calls himself or herself a scientist or an engineer, regardless of the type of scientist or engineer and regardless of the area of scholarship. Thus, introducing to the court a member of either of these two professions as a potential expert witness presents neither problem nor question to the jury.

The scientist or engineer new to the field of testifying in legal cases must decide if he or she really wants to be subjected to the ordeal of the cut and thrust of the courtroom proceedings. He or she should do some careful thinking about the realities of a courtroom battle because the scientist or engineer will, most likely, have to stand up to potentially abusive attacks. The outcome of a trial is more often based on the expert's credibility than that of any other witness.

However, the scientist and engineer must remember that the area of scholarship of an attorney is the *law* and the law is constructed from *words*. Even the most knowledgeable witness who is willing to sit back in the witness chair and pontificate (to show how knowledgeable he or she really is) will be led down the garden path by any attorney worth his or her salt until contradictions in the expert's testimony are evident and the opinion of the expert ends up on the compost heap of disbelief!

However, before delving into the body of the text of this book, it is necessary to define an expert and the general rules of court procedure.

1.2 AN EXPERT WITNESS

The term *expert witness* is a loosely used term and, in fact (in the context of this book), a scientist or engineer is not an expert witness until the time when a court of law (the judge) confers such recognition. When this is done, the expert is permitted to render opinions as evidence and the opposing disputant (in the form of opposing counsel) may not object to the witness testimony on the grounds that it is an opinion.

On a more definitive level, the term expert refers to a person who has a defined level of expertise in his or her chosen field of scholarship by virtue of education, training, skill, or experience and has knowledge beyond that of the typical person. Thus, an expert witness is a witness, who by virtue of education, training, skill, or experience, is believed to have knowledge in a particular subject beyond that of the average person. This is sufficient knowledge so that others may officially (and legally) rely upon the witness's specialized (scientific, technical, or other) opinion about an evidence or fact issue within the scope of the expert's expertise, referred to as the expert opinion, as an assistance to the judge and jury.

On the other hand, to some readers, an expert may seem to be a person who has a briefcase or a laptop computer or some other symbol of the electronic age and is at least fifty miles away from home. The first definition is applicable to the subject matter of this book. The second definition is not the focus of this book but still deserves mention in the following pages.

To further define an expert, the term *chosen field of scholarship* refers to any area of science, technology, business, religion, or the arts that requires an opinion. The expert, of course, is the person who is willing to express an opinion related to his or her area of scholarship verbally (as in court) or in writing (as in the form of an expert report or letter of opinion). An expert witness serves justice. He or she appears in court or before a tribunal to assist the trier of fact (the judge, the jury, the arbitrator) by explaining complex technical issues and conveying expert opinions and conclusions as to the importance of these issues as they relate to the matters under dispute.

With reference to the expert witness who plays a role in a court case, there are two main types of expert witness: (1) the *fact witness* (*eyewitness, lay witness*) and (2) the *expert witness* (*technical witness*). In both examples, the lay witness or the expert witness must present the truth to the court.

The fact witness is called to the witness stand to tell the court what he or she has personally seen or heard. In such cases, the testimony of the fact witness is limited to what this witness has actually seen or heard (i.e., to facts derived from a first-hand basis related to the matter at hand). The fact witness can therefore only testify on matters related to knowledge of certain limited facts. The fact witness cannot interpret or describe what others did, saw, or thought. In addition, the fact witness cannot express an opinion on the subject in the trial. The fact witness can tell the court (the judge, in the case of a bench trial, or the jury, in the case of a jury trial) what he or she has experienced, but this witness cannot opine (express an opinion) on what he or she thinks on the matter related to the trial.

In contrast, an expert witness (technical witness) can opine (express an opinion) within the scope of his or her experience as well as about the facts or opinions that have been presented by others in previous testimony or in expert reports. As broad as this may seem, there is, nevertheless, a boundary for the scientist and engineer who is designated as an expert for the matter at hand. In general, the expert should not introduce issues that are marginally peripheral or unrelated and outside his or her scope of designated expertise.

Although the decision to allow such unrelated or peripheral testimony is at the discretion of the judge, there is one hard and undeniable fact: the expert must present the truth to the court in a manner that is free of errors and embellishments. The testimony must be unencumbered by any previous ties to a person, to an organization, or to any party line and must not be perceived to be biased. In short, the expert must not be an advocate for one side or the other.

Above all, a scientist or engineer who is being retained as an expert must be of good character, honest, and committed to good science, among many other things (Table 1.1). The expert must also be accessible to the attorney who hires the expert. If the expert is too busy, the attorney must be so advised early on. In addition, the expert must be an expert in communication skills as well as the science of the case. Thus, to be an expert, a scientist or engineer must have a combination of special knowledge, skill, experience, training, and education to qualify as an expert on the subject in dispute.

The scientist or engineer can be requested to testify to his or her knowledge of any one or more areas within his or her technical field of expertise. It is doubtful that any one person could be an expert in every area of his or her profession (and attorneys are cautioned to guard against such claims), so it is important that the scientist or engineer admit to specialization in certain specific areas of his or her discipline. Examples would be a chemist who specializes in process chemistry

TABLE 1.1
Requirements for an Expert Witness

The Expert Must
Recognize that his or her task is to present the truth to the court
Tell the truth
Ensure that personal and professional background is beyond reproach
Present testimony that pertains to a subject whose technology is beyond common experience
Use authentic references in reports
Not use junk science
Present testimony based on material that is known to be valid
Present testimony that is based on special knowledge, experience, education, and skill
Present testimony that is limited to a specific area of his or her knowledge
Present testimony that demonstrates the validity of his or her analysis of the facts
Present testimony that allows valid conclusions to be drawn from the facts
Speak understandably
Speak authoritatively
Answer the questions simply, preferably with a *yes* or *no*
Only elaborate on an answer when asked
Evaluate the strengths and weaknesses of the argument
Adhere to the required confidentiality
Maintain good relationships with the attorney
Be courteous to opposing counsel and all others involved in the case
The Expert Must Not
Take on an assignment for the money only
Place himself or herself in a position of conflict of interest
Lie or give false information
Embellish a résumé or use résumé *fluff*
Miss any deadlines for written reports
Be disrespectful to the court
Speak in hushed or inaudible tones
Pontificate
Attempt to interpret the law
Allow a disrespectful or abusive attorney to intimidate him or her
Throw away files until advised by counsel to do so

and an engineer who specializes in reactor engineering but has only marginal knowledge or experience of petroleum refining and petrochemical techniques.

It is the joint responsibility of the expert and the attorney to determine in which area or areas the scientist or engineer will qualify as an expert. The lack of expertise in a given area will undoubtedly be revealed and probed in depth by the opposing counsel during cross-examination.

Often, an expert with practical experience is more acceptable to jurors and thus more favorably received by the court than one who deals only in the realm of academic theoretical evaluation. A question currently in dispute is whether the expert must have some publications in a refereed (peer-reviewed) journal.

The expert's testimony must pertain to a subject that is sufficiently beyond common experience. The information given by the expert must be of benefit to the judge and jury, if it is assumed that neither the former nor the latter should generally be expected to have this knowledge. It is also expected that a reasonably educated jury should not have this specific technical knowledge.

The prospective expert should realize that some lawyers will try various means to keep an opposing expert out of court. There may be attempts to prevent the opposite side from hiring the expert. For example, one way to keep an expert from helping the opposition is to sign him or her up as a consultant, giving him or her very little work to do while giving him or her confidential information. Ethically, the expert with this information cannot discuss the case with others and is therefore never available for the opposite side. Another tactic is to engage an expert as an expert and have a written contract but then never use him or her. In either case, the expert is safely removed from action.

The scientist or engineer must, at the time of the initial contact, inform the attorney who is interested in engaging him or her as an expert that this is a preliminary interview, and only general terms can be discussed. The scientist or engineer must discuss matters at issue in the case, but privileged or confidential information should not be discussed.

If the other side calls, it is important for the potential expert to make it clear that he or she has already been contacted by an opposing attorney. The decision to accept or not to accept an assignment is a personal choice.

It cannot be overemphasized that the expert in court is not an advocate but, rather, acts as a balanced expert and presents the truth to the court. The expert must be objective and must not be an advocate. Moreover, even though the expert's views may change somewhat as he or she learns more about the facts of a case, the expert must never alter his or her thinking or written opinion just to please the attorney.

In many cases, the alleged damages are based on technical issues that can only be explained by the technical expert. The time factor also is important. In fact, some defendants may see litigation as a means for delaying payment or paying less than is claimed, and they can continue to use their money to make more money, thereby offsetting some of the damages. Those with the most money are generally the ones who can afford the best legal counsel, the best experts, the most waiting time, and the high fees needed for the various proceedings used to wear down the opposition into a submissive role.

However, seldom does litigation run its full course insofar as the majority (often in excess of eighty percent) of all suits (federal and state) are settled out of court. For those cases that do proceed to trial, the full course is not necessarily attained when the judge or jury delivers a verdict. The verdict can be appealed, and the appellate decision can also be appealed. In fact, the full course may not be run until the matter is resolved by the Supreme Court of the United States.

Finally, the expert's testimony must be based on material of a type that an expert may reasonably rely on to form an opinion. An expert is not needed to state facts to which anyone can attest. The use of an expert is necessary to go beyond the facts of the case and present an opinion to the jury based an analysis of the facts.

1.3 TYPES OF EXPERTS

Since the U.S. Supreme Court's decision in *Daubert v. Merrell Dow Pharmaceuticals, Inc.*, which clarified the standards for admitting new scientific evidence, there has been extensive comment on the expanded use of expert evidence in the courts. In particular, many comments have focused on the fields of mass tort litigation and criminal law. No less dramatic is the parallel increase in the use of expert evidence in federal civil rights litigation. Beginning with the case of *Brown v. Board of Education,* this trend is reflected in the diverse issues involving expert proof, often novel in nature, that have regularly appeared in the Supreme Court's civil rights decisions.

The proliferation of new types of expert evidence in federal civil rights cases has not gone unchallenged. Disputes regarding the admissibility of expert evidence are now routine, and the frequency with which issues of expert evidence arise in civil rights cases is unlikely to diminish for a number of reasons. First, the technocratic nature of our society, including its increasing reliance on technological means to regulate human behavior in areas such as employment and law enforcement, generates many civil rights cases. Second, case law and statutes have recognized new claims, such as exposure to unreasonable health risk from tobacco smoke or discrimination due to disability, that ordinarily require consideration of expert evidence.

Litigation of the type referenced in the current context usually involves complex technical issues. An attorney handling a computer matter typically cannot assess the technical aspects of his case. Judges and juries ordinarily lack the background to evaluate technical evidence. Arbitrators, who may have the technical expertise required, are usually lay people, unable to understand the legal issues in the case. The evaluation and presentation of a case obviously require the use of technical experts who, among other things, can reduce complex technical concepts to terms readily understandable by nontechnical people.

The selection of a bad, poor, or mediocre expert witness or the improper use of a good expert can, to paraphrase a well-known and very pertinent phrase, cause defeat to be snatched from the jaws of victory. In short, the expert should be chosen not only based on his or her qualification for the technical aspects of the dispute but also on the role that the attorney sees for that particular expert in the dispute and the credibility of the expert before the judge and jury.

Thus, before proceeding, it is necessary to define the types of expert: (1) the *nontestifying expert*, (2) the *testifying expert*, (3) the *in-house expert*, and (4) the *court-appointed* expert.

1.3.1 The Nontestifying Expert

The *nontestifying expert* should be as knowledgeable as the expert; it is his or her role to provide advice to the attorney. The nontestifying expert may be hired to help a potential claimant evaluate a case in terms of the scientific and engineering principles (in the current context).

A nontestifying expert does not generally need the qualities sought in his or her testifying counterpart. This is not to deny that those qualities are not welcomed; they simply are not necessary for his or her function. The demeanor and ability of the nontestifying expert to connect with a judge or judge and jury are not essential. Instead, his or her knowledge and experience are of more importance, and the ability to analyze the data and help assist the attorney at depositions and at trial is paramount through candid evaluation of the issues in the case.

As a nontestifying expert, the expert actually becomes part of the attorney's legal team insofar as he or she is an advisor to the attorney. Because he or she is retained to assist in litigation, the expert becomes subject to and protected by various privileges and immunities. Accordingly, conversations between the nontestifying expert and the attorney are *generally* protected. Moreover, the materials prepared for the lawsuit by the nontestifying expert *may be* immune from production. Thus, the advice given by the nontestifying expert to the attorney may be privileged (opinions differ on this matter or—no pun intended—the jury is still out on the question of privilege) and not subject to discovery. To assist in ensuring privilege, written reports, notes, comments, and invoices should contain word headings such as "Privileged and Confidential—Prepared at the Request of Counsel." Such notes are the attorney's work product; however, privilege is lost if any of these work products are given to anyone other than the attorney.

Included in the tasks of the nontestifying expert are assisting the attorney with questions to ask the opposing expert during the deposition and framing technical questions to ask prospective jurors. Advice can be given to the attorney about the type of juror who can understand the nature of the expert's testimony; it is desirable to get at least one juror who will understand the science of the case. This kind of expert opinion may be protected from discovery, and if the expert finds factual data that are against the potential defendant, the data can remain privileged (i.e., not divulged to the defendant). This privilege is similar to the work product protected by the attorney–client privilege.

The information collected by a nontestifying expert may be protected from discovery under Rule 26(b)(4)(B) of the Federal Rules of Civil Procedure, which bars parties from obtaining discovery of facts known or opinions held by an expert who has been retained or specially employed by another party in anticipation of litigation or preparation for trial and who is not expected to be called as a witness at trial, except upon a showing of exceptional circumstances. Prior to the enactment of Rule 26(b)(4)(B), courts relied on two different theories to prevent discovery of information obtained by nontestifying experts. Some courts suggested that this information was protected only if it fell within the purview of the attorney work product privilege. A number of other courts held that discovery of nontestifying experts should be prohibited on the basis of fairness insofar as it would be inherently unfair for one party to retain an expert, finance the expert's investigation, and then be forced to turn the expert's facts and opinions over to an opponent.

However, attorneys and experts would do well to remember that, when one party or the other decides to retain or appoint a nontestifying expert, the court may find that sufficiently exceptional circumstances exist to warrant disclosure of the expert's findings (http://www.rkmc.com/Discovery-Ordered-from-Consulting-Non-Testifying-Expert.htm). On the other hand, the attempt by an attorney to call to the witness stand a scientist or engineer who has previously been declared a nontestifying expert may be declined (http://www.gsbca2.gsa.gov/oldappeals/w13125a.txt).

1.3.2 The Testifying Expert

If the witness is required to testify in court, the expert becomes a *testifying expert* and privilege is no longer protected. The expert witness's identity and all documents used to prepare the testimony will become discoverable. Usually, an experienced lawyer will advise the expert not to take notes on documents because all of the notes will be available to the other party.

The role of the testifying expert is very different from the role of the nontestifying expert. Although he or she is often viewed as a member of the legal team, the law treats him or her differently:

The existence of the testifying expert must be disclosed to the opposing counsel.

The expert is generally required to prepare and sign a report containing a complete statement of opinions to which he or she will testify and the basis for those opinions.

The testifying expert must disclose all of the facts, data, and documents upon which he or she relied to form these opinions and to prepare the report.

The testifying expert must be prepared to inform opposing counsel of his or her qualifications and to provide a list of the cases in which he or she has recently testified.

The testifying expert must be prepared to disclose the compensation for serving as a testifying expert.

Because of the distinct roles they play, potential experts fill these needs differently. A testifying expert needs to be able to articulate his or her thoughts and opinions clearly and simply. The expert needs to have the skills of a teacher and mentor to the judge and the jury. He or she must be patient and be able to think through the questions posed by the adversary. The expert must be able to speak with confidence but not appear self-centered as well as be a good listener and appear thoughtful. In short, the testifying expert does not need to be the smartest scientist or engineer in the relevant field but must be competent and clear.

The testifying expert must be a scientist or engineer who has special knowledge, experience, education, and skill to qualify as an expert. In addition, he or she must be able to present testimony that is (1) of benefit to the court in adjudicating the case, (2) limited to a specific area of knowledge related to the case, (3) related to a subject in which the technology is beyond common experience, and (4) based on admissible evidence that the expert may rely on to form an opinion.

The expert may discuss the opinion of another expert (even if that opinion is not admissible in evidence, subject to guidance from the attorney and permission from the judge) but cannot use only the conclusions of other scientists. Furthermore, even though circumstantial evidence can be used for an opinion, the expert must be careful to establish the foundation for such an opinion and endure the avoidance of conflicts (he said/she said and the like).

Subject to the judge's permission (obtained through the attorney and in the court), it is often permissible for an expert to give an opinion on hearsay testimony, which is normally excluded from court hearings. This hearsay testimony must be of a type usually relied upon by other experts in the field in which this expert is testifying; thus, there is some leeway for the expert to express an opinion on the case. For example, unpublished information may be relied upon by the expert, but it is advisable that an expert worth his or her salt not rely upon unpublished information to any great extent.

Many judges and juries are suspicious of hearsay evidence or unpublished data. The frequent use of such evidence (or data) can cast doubts upon what otherwise may be an excellent testimony. The difficulty is whether the court thinks the testifying expert has actually relied on the opinion of the other expert or whether the expert affirms that the other expert agrees with the testifying expert's opinion. A major problem occurs when the expert includes in his or her opinion information from a conversation with another expert.

In some cases, an expert may testify that the discussion or even the opinion he or she gives, although counter to mainstream science, is derived from actual facts. The complication is that some theories are sound, whereas others are *junk science* and are not based on any form of logic or reliable data. It is often difficult for a lay group like a jury to distinguish between such testimonies; failing to understand the real facts behind the testimony, the jury (understandably) will be suspicious. The testifying expert will have as much credibility as the carpetbagger of decades past who was clothed in the loudly colored plaid suit and white shoes.

Credibility is also undermined when the testifying expert uses noncredible scientific or engineering testimony, studies, or opinions. It has become all too common for expert testimony or fringe studies that are outside the scope or limits of mainstream scientific or engineering views to be presented to juries as valid evidence from which conclusions may be drawn. The use of such invalid scientific or engineering evidence (*junk science, junk engineering*) has resulted in finding of causation that cannot be justified or understood from the standpoint of the current state of credible scientific and engineering knowledge.

The testifying expert must remember that the judge or the jury will have to decide between the different conclusions presented by opposing expert witnesses. The jury may not be able to differentiate between the scientist with credentials related to the topic before the court and the scientist who may also have impressive credentials, not necessarily in the same discipline, who presents an unconventional interpretation of the data.

Unfortunately, there are a number of scientists and engineers without formal training in the subdiscipline before the court who are, nonetheless, being represented as experts in that subdiscipline. A few approach the witness stand with an air of authority and then present theories and conclusions contrary to accepted dogma. In many cases junk science is expounded and the result is a battle of the personalities and appearances of the testifying experts.

Therefore, the expert should be able to explain some basis of distinguishing real from junk science. One can conclude that a theory is junk science if the pet hypothesis cannot be repeated and verified. The expert should present to the court and the jury whether or not the hypothesis has been published in a peer-reviewed journal. He or she must explain that a peer-reviewed article has the most likely chance of having few, if any (there are no absolutes), flaws in the scientific or engineering methodology. Moreover, the expert should be able to testify whether or not the hypothesis has widespread acceptance, though this is not the sole criterion for acceptance as good versus junk science.

The testifying expert must also be aware of the latest advancements in his or her scientific and engineering disciplines. The expert must recognize that scientific and engineering disciplines are dynamic insofar as ideas and theories can change as more experimental data are added for consideration.

In contrast, the reason that the expert has been called to testify is the need for the dispute to be placed before the legal system to be resolved within a finite time period of time. Thus, it is possible that much science and engineering will be decided (at least, one hopes for the time being) by the legal system and not by the body of data contained within the scientific and engineering disciplines.

In the end, the testifying expert must demonstrate the validity of his or her analysis of the facts and the conclusions drawn from these facts.

1.3.3 The In-House Expert

The *in-house expert* is, for the purposes of this text, a scientist or engineer employed by the plaintiff or by the defendant who is designated as a testifying expert in the case. The in-house expert represents an easily available source of expertise to which the other party does not have access. However, prior to using an in-house scientist or engineer as an expert witness, the attorney must first check to see if the employment contract of that scientist or engineer permits him or her to testify in a court case.

Using in-house experts is often a preferred modus operandi if the company employs someone with some scientific or engineering training or education. A judge and jury may expect to hear from a company representative about the safety and testing of the product under question in the case. In-house experts may also be useful in assisting outside experts, especially if the product is complex.

There are many advantages as well as disadvantages to using an in-house expert to testify for either side. The advantages are that this expert (1) knows the company, (2) knows the intimate details of the internal company workings, (3) can help the attorney become acquainted with the industry environment, and (4) knows the history of any laboratory procedures and protocols that might be relevant to the case. The disadvantages of using an in-house expert include that (1) the jury can be told the in-house expert must testify to keep his or her job; (2) opposing counsel can allude to the pressure on the in-house expert to testify, regardless of the facts; (3) the expert can give away trade secrets under cross-examination; (4) the expert can be too critical of the company practices, adversely affecting the company's position in the case; and (5) the in-house expert may suffer from the stigma of being the defendant's paid spokesperson.

The implication is always that, although the in-house expert is under oath, the company expects him or her to be willing to commit perjury. Perhaps the wisest strategy in any case is to use an in-house expert and an outside expert who can support the testimony of the in-house expert.

It is conceivable that the in-house expert may be declared a nontestifying expert. However, courts are divided on whether the discovery of in-house non-testifying experts is protected by Rule 26(b)(4)(B) of the Federal Rules of Civil Procedure. The dispute centers on the language that an expert must be *retained* or *specially employed* in anticipation of litigation or preparation for trial.

1.3.4 The Court-Appointed Expert

Under certain circumstances, but rather infrequently, the court may appoint its own expert. Often the need for such an expert is not recognized by the court. A judge has the authority to select specialists such as translators and speech handicap assistants thus, it is logical for a judge to appoint a scientist or engineer, if necessary, as well.

The opposing parties may not be in favor of this strategy, and they may indeed be strongly opposed. If this is the case, they will not make suggestions as to who should be appointed as the expert or participate in any way in the appointment. Sometimes, however, this special expert is selected with the consent of both parties. Once designated, the expert must be notified in writing, and his or her specific duties must be spelled out. The *court-appointed expert* can greatly influence the outcome of litigation and at times facilitate settlement before a case goes to trial.

The court-appointed expert can also be most helpful to the judge in pretrial hearings to help screen scientific testimony. The jury looks on the court-appointed expert as a truly neutral expert, who has much knowledge in the field in question and no regard for the interests of either party to the litigation. This expert can break an evidence stalemate. However, the court-appointed expert is subject to cross-examination by attorneys from both sides.

1.4 TYPES OF LITIGATION

Many expert witnesses do not know the differences between the various types of litigation. It is good to be advised of such types of litigation because this may make a considerable difference to the scientist's or engineer's response when first contacted by the attorney.

Briefly, a controversy before a court or a *lawsuit* is commonly referred to as *litigation*. If it is not settled by agreement between the parties, it will eventually be heard and decided by a judge or jury in a court. Litigation is one way that people and companies resolve disputes arising out of an infinite variety of factual circumstances.

The term *litigation* is sometimes used to distinguish lawsuits from *alternate dispute resolution* methods such as *arbitration,* in which a private arbitrator would make the decision, or *mediation,* which is a type of structured meeting with the parties and an independent third party who works to help them fashion an agreement among themselves. Litigation occurs when there is a dispute between two or more parties. In some cases the disputants will attempt to reach accord on their own or with the assistance of their attorneys. When this fails, the dispute goes into litigation at the pleadings stage.

An expert may start serving one of the disputants almost immediately, although this usually happens in the later stages of the process. This in itself is an important consideration for the scientist or engineer in deciding whether or not he or she will accept the assignment.

The research conducted by an expert is generally the same no matter what procedures are used for resolution of the dispute. However, the types of litigation vary, which may require differences in presentation of the facts as well as the types of reports required by the dispute-resolving body. Many forms of dispute resolution procedures are used. Some provide for nonbinding discussions between the disputants; others require binding settlement of the dispute. In all cases, the expert can perform a valuable service by presenting impartial testimony and conclusions based on an appropriate degree of professional inquiry and research.

Thus, a few lines of description for the types of litigation can be helpful for the blossoming expert.

1.4.1 HEARINGS

A *hearing* is a legal proceeding before a judge or court referee, other than an actual formal trial, and is any proceeding in which the disputing parties have the opportunity to present evidence or testimony to the court or fact-finder. A hearing is usually shorter than a trial but still may result in a final order.

Hearings are conducted for various reasons; for the most part, however, they are held to resolve disputes related to contract issues or to resolve allegations made against licenses. More often than not, experts are required to validate the legitimacy of the positions of the disputants. Hearings may also be held to gather information prior to making a decision or recommendation. Typical hearings may be conducted to resolve issues related to land use and water and mineral rights to reach a fair value on the land use or water use.

Hearings resemble trials in that they ordinarily are held publicly and involve opposing parties. They differ from trials in that they usually feature more relaxed standards of evidence and procedure and take place in a variety of settings before a broader range of authorities (judges, examiners, and lawmakers). Hearings fall into three broad categories: *judicial, administrative,* and *legislative.*

Judicial hearings are tailored to suit the issue at hand and the appropriate stage at which a legal proceeding stands. They take place prior to a trial in both civil and criminal cases. *Ex parte hearings* provide a forum for only one side of a dispute, as in the case of a temporary restraining order (TRO), whereas *adversary hearings* involve both parties. *Preliminary hearings,* also called preliminary examinations, are conducted when a person has been charged with a crime. Held before a magistrate or judge, a preliminary hearing is used to determine whether the evidence is sufficient to justify detaining the accused or discharging the accused on bail. Closely related are *detention hearings,* which can also determine whether to detain a juvenile. *Suppression hearings* take place before trial at the request of an attorney seeking to have illegally obtained or irrelevant evidence kept out of the trial.

Administrative hearings cover matters of rule making and the adjudication of individual cases and are conducted by state and federal agencies. Rule-making hearings evaluate and determine appropriate regulations, and adjudicatory hearings try matters of fact in individual cases. The former are commonly used to garner opinion on matters that affect the public, such as when the Environmental Protection Agency (EPA) considers changing its rules. The latter commonly take place when an individual is charged with violating rules that come under the agency's jurisdiction—for example, violating a pollution regulation of the EPA or, if incarcerated, violating behavior standards set for prisoners by the Department of Corrections.

The degree of formality required of an administrative hearing is determined by the interest. The greater that interest, the more formal the hearing will be. Notably, rules limiting the admissibility of evidence are looser in administrative hearings than in trials. Adjudicatory hearings can admit, for example, hearsay (a statement by a witness who does not appear in person, offered by a third party who does appear) that generally would not be permitted at trial.

Legislative hearings occur at both the federal and state levels and are generally conducted to find facts and survey public opinion. They encompass a wide range of issues relevant to law, government, society, and public policy. Legislative hearings occur in state legislatures and in the U.S. Congress and are a function of legislative committees. They are commonly public events held whenever a lawmaking body is contemplating a change in law, during which advocates and opponents air their views. Not all legislative hearings consider changes in legislation; some examine allegations of wrongdoing. Although lawmaking bodies do not have a judicial function, they retain the power to discipline their members, a key function of state and federal ethics committees.

1.4.2 Mediation

Mediation is an informal process that allows people who are involved in a dispute to resolve their dispute without any legal proceedings. It, together with arbitration, is a type of *alternate dispute resolution*. However, mediation is different from arbitration.

Mediation is a favored alternative to litigation and is one of the most preferred forms of *alternative dispute resolution* (ADR) for avoiding or settling litigation. In its simplest terms, mediation is a nonbinding negotiation between adversaries that is conducted with the assistance of, and often through, an experienced, neutral third party. Except under some court-mandated programs, mediation is a consensual effort to which both parties must agree. Mediation is often employed after it becomes apparent that direct negotiation between disputants or adversaries will not resolve the dispute efficiently.

Some differences between mediation and litigation include:

- Mediation
  Time: hours or days
  Cost: hundreds to thousands of dollars
  Privacy: total
  Formality: low to median
  Determination: enforceable
  Right of appeal: none
- Litigation
  Time: months or years
  Cost: thousands to hundreds of thousands of dollars
  Privacy: none (media may be present)
  Formality: high
  Determination: enforceable
  Right of appeal: yes

An attractive aspect of mediation is that it can be tailored to suit the needs of each individual dispute. Competent mediators come from a wide variety of professions and employ different styles. Important factors in choosing a mediator include experience, reputation, educational credentials, mediation training, apprenticeships, gender, age, cultural background, knowledge of a particular field, and accreditation by mediation organizations or courts.

The mediator can play a low-key and conciliatory role or take on a more proactive role by making suggestions and probing for convergent interests. The parties can also decide to convert the mediation into an arbitration proceeding, granting the mediator the power to issue a binding decision. In general, the mediator gains (or attempts to gain) the confidence of both sides of the dispute to probe for *fallback* positions. He or she will then strive to attain the middle ground that both sides can accept.

Mediation does not always rely on specific points of law. Problems can be solved by the disputants looking to the future instead of finding fault or blame. In contrast, the courts make judgments based upon the law and the rules of proceedings limit what can be considered. In fact, a court may be unable to address the underlying issues in a dispute.

Most mediations end with a written agreement that outlines all the details of the settlement. Written in plain language (instead of legalese), the settlement is specific in the commitments of the participants and may include steps to follow if similar problems come up in the future. However, mediation agreements should be reviewed by an independent attorney or other expert before they are signed.

The importance of a nonbiased expert in mediation proceedings is essential; an expert who acts as an advocate can have an adverse effect on the outcome of mediation proceedings.

In summary, mediation gives disputants or opponents a quick way to work out their differences while addressing everyone's needs and interests. The privacy of mediation can make it easier for people to discuss emotional matters. The decisions reached are not imposed on the disputants by a judge, and compliance is much higher in mediated cases than litigated cases. In addition, an attractive aspect (although not the only reason for choosing mediation) is that mediation is always less expensive than going to trial and services are sometimes available free of charge.

1.4.3 ARBITRATION

Arbitration is the best known alternative to civil litigation and permits disputes to be resolved quickly. However, the time and cost can be as onerous as those that arise because of civil litigation.

Instead of a judge and a jury, an arbitrator or panel of arbitrators is used and selected by the disputants through a process of elimination. The arbitrator's background should give him or her a general understanding of the issues involved, which may modify the role of the expert as far as hisor her testimony is concerned.

Nonbinding arbitration (or *advisory arbitration*) is a process that resembles arbitration in that the parties agree to submit a dispute to an arbitrator but agree that the award is merely advisory. The parties need not follow it and each party is free to pursue an independent action, such as a lawsuit in a court of law.

On the other hand, *judicial arbitration* is a procedure under some state laws (including California) that is neither judicial nor arbitration. It is a dispute-resolution technique that occurs after a lawsuit has been filed and before a trial is held. The lawyers for litigating parties each present their side of the case to a third (and independent) lawyer, who then gives his or her opinion on who would win and how much the loser would pay. The parties can accept this opinion or continue their litigation. If one side does not accept the opinion, there are certain consequences if that side fails to do better at trial.

Unlike litigation, arbitration often eliminates or significantly modifies the discovery process by minimizing each disputant's ability to examine documents and relative positions. However, it is not unusual that both arbitration and litigation may be required to resolve a dispute, and the decision from one may be the opposite of the decision from the other.

Although it may occasionally be appropriate that some disputes arising out of a contract should be resolved by arbitration while others are dealt with in the courts, that is usually not the case. Arbitration may be appropriate in the following situations:

- where court procedures would be lengthy and expensive and the parties are able to agree to their procedures, such as: (1) the level of representation in the arbitration; (2) the extent to which oral argument may be permitted; (3) subject to the tribunal's availability, the time of the hearing; and (4) the language of proceedings and documents to be used in the course of the proceedings;
- where expertise in the subject matter of the dispute is an important attribute of the person who is to decide the issues; and
- where confidentiality is required.

Whatever the situation, the parties negotiating a dispute resolution provision will each try to have the dispute resolved in the forum that favors their individual interests. Furthermore, the choice between arbitration and litigation in the courts is not a choice that can be made in a vacuum. The identities of the parties, the location of assets, the nature of the dispute (potential or actual), and the courts that might otherwise have jurisdiction are only a few of the many factors that may have to be considered.

Finally, lower cost is claimed (but not often substantiated) to be one of the most important benefits arbitration has over litigation. A lawsuit can cost upwards of $50,000; arbitration may cost only a tenth as much.

A drawback to arbitration is that, in many jurisdictions, the arbitrator (or panel of arbitrators) is not required to file a written opinion. Often, the arbitrator hands down a one-paragraph decision, awarding a decision in favor of one side or the other.

1.4.4 Mediation/Arbitration

Mediation/arbitration, also known as *binding mediation,* has been used to resolve labor disputes. In essence, such disputes require a mediator familiar with the arbitration issues. The mediator gathers testimony and information about the nature of the dispute and encourages mediation. If mediation fails, the mediator/arbitrator may then impose binding arbitration. At such times, each disputant may select his or her own expert, although the mediator/arbitrator can resolve the dispute by also retaining his or her own expert.

1.4.5 Civil Litigation

Civil litigation is probably the best known but least understood resolution procedure. In civil litigation, the dispute is between two or more *parties* (individuals, businesses, or government agencies). Most often, the result is an award of money to be paid by one party to the other. The judgment is imposed to make the aggrieved person whole for the harm that has been caused by the other. A judgment in a civil matter does not include the imposition of a criminal sentence.

The civil procedure can move along well-worn pathways, as mandated by laws that differ from jurisdiction to jurisdiction. Decisions relative to highly complex technical issues are made by laypersons who often gauge credibility (including the credibility) of the expert by factors that may not always be related to the matters under dispute. For example, the comfort or discomfort of the expert witness, the manner in which he or she is dressed, and his or her body language may all play a role in the decision-making process. But no matter what the decision, it may be subject to one or more appeals or retrials.

The time involved in civil litigation can be extensive, even though court calendars are crowded. In fact, it may not be unreasonable to expect several years to pass from the date of filing the suit before a final verdict is obtained.

The expense involved can be extraordinarily high and, in some cases, may exceed the total amount at issue. This may be to (or not to) establish a precedent or to satisfy the code of honor of the plaintiff. It is for cost reasons that many disputes end in out-of-court settlements.

The rules of civil procedures are different from those of criminal procedures because proceedings are different.

1.4.6 Criminal Litigation

In criminal matters, action is taken by federal, state, or local government agencies against an individual for a violation of the law. A criminal matter can result in a sentence such as a fine, probation, or time in jail. The sentence is imposed upon a defendant who pleads or is found guilty to keep him or her from acting in the same manner in the future and also to deter others from acting in a similar manner. Because a criminal matter can result in the "state" taking away a person's freedom, there are additional constitutional protections built into the rules of criminal procedure.

1.4.7 Private Litigation

Private litigation requires both sides to agree to resolve their differences in a court that they create and for which they pay. In such cases, a mutually respected individual (such as a retired judge) is chosen to act as the judge. The usual rules of civil litigation are followed, except that the trial is a bench trial in which the judge acts as judge and jury. The advantage of this approach is that the decision can be reached quickly and is not subject to the usual court load of other cases that must be decided first. In addition, when major amounts of money are at stake, the money savings are extremely high.

1.4.8 Class Action

A class action is a legal procedure used to handle a lawsuit in which a large number of people have been injured (mentally, physically, and financially) by a common act or set of actions. A class action lawsuit is a civil lawsuit that is brought by one person or a few people on behalf of a larger group of people who have suffered similar harm or have a similar claim. Class actions can be brought in federal or state courts, though a federal law, the Class Action Fairness Act of 2005, makes it easier for defendants to move class action lawsuits from state to federal courts.

Class actions are generally used when too many people have been affected by the subject of the claim for each of them to file a separate lawsuit. For example, class action suits are frequently used for claims of injury from hazardous products, including pharmaceutical products and dangerous drugs. Class actions are also frequently to stop illegal or harmful practices like oil spills, manufacturing pollution, or violations of state or federal constitutional protections.

A lawsuit becomes a class action when one or more plaintiffs, often called *lead plaintiffs,* file a lawsuit claiming harm or damages. The plaintiffs can then ask the court to certify the case as a class action. To have the case certified, the lead plaintiffs and class action attorneys must show that the case meets several criteria:

There is a legal claim against the defendant.

There is a significantly large group of people who have been injured in a similar way, and the cases of members of the class involve similar issues of fact and law as the case of the lead plaintiffs. Class certification might be denied, for example, if people have suffered different kinds of side effects from a defective drug. The differences in injury would require different evidence for many class members.

The lead plaintiff is typical of the class members and has a reasonable plan and the ability to represent the class adequately. The lead plaintiff must also have no conflict with other class members. A lead plaintiff who seeks money damages for himself or herself but is willing to agree to coupons for all the rest of the class is probably not adequately representing the class.

If a lawsuit is certified as a class action, the court will order that the class of people affected be notified, which is done through direct mailings as well as through the media and Internet. Class membership is automatic in all but a very few cases. All individuals affected by the action or product complained of will be part of the case unless they choose to withdraw.

Class members do not take part in the case directly unless they have evidence to offer, and they are not involved in the decision of whether or not to accept a settlement offer. The lead plaintiff consults with the class action attorneys to plan the strategy of the case and to accept or reject settlement offers. Other class members have only the option of opting out of the settlement.

The court decides how to divide any recovery at the end of a class action suit. The attorneys are given costs and fees, often calculated as a percentage of the entire recovery, the lead plaintiffs receive an amount partly determined by their participation in the lawsuit, and the rest of the recovery is divided among the class members.

1.4.9 Appeals

In its broadest sense, an appeal is a formal request that a higher body—typically a higher court—review the action, procedure, or decision of a lower court, administrative agency, or other body. The disputant who loses the dispute or did not get all the relief he or she sought usually makes an appeal. If both disputants are dissatisfied with the decision, each may appeal part of the decision.

1.5 RULES OF THE COURT VERSUS RULES OF THE LABORATORY

In many cases as much as ninety-five percent or more of all work performed by scientists and engineers who appear as expert witnesses occurs outside the courtroom or recognized dispute tribunals. The vast majority of cases may never reach trial insofar as a settlement is achieved beforehand, often as a result of the expert's findings or as a result of the plaintiffs or the defendants seeking to avoid the enormous costs of a trial and the uncertainty of its outcome. Nevertheless, standards of behavior must be observed, and the scientist or engineer who tends to see an appointment as an expert witness as being a lucrative line of employment must understand the rules and standards of behavior required of the expert.

Scientists and engineers enjoy voicing their opinions about any item that comes to mind. Such opinions will be brought forth in the laboratory, during the coffee break, and during the lunch break and are verbally presented to anyone who is willing or unwilling to listen. These worthy individuals (the scientists and engineers) are encouraged to do this by their colleagues, by chance acquaintances, or often by perfect strangers. Often these opinions leave their respective fields of expertise and extend into areas that have no bearing whatsoever on their respective historical paths of education and knowledge.

Usually, the scientists and engineers who make good expert witnesses are sensible persons who maintain a dignified silence and learn from the wisdom and the folly of others. The first category (i.e., learning from the wisdom of others) may warrant some degree of verbal participation in the conversation or discussion. The second category (i.e., learning from the folly of others) does not warrant any response from the listener. Merely walking away when one has heard enough will suffice.

Beyond the laboratory, the "coffee klatsch," and the lunch bunch, the rules are different. It is not how knowledgeable a person may appear to be; rather, an expert in a case is only an expert if so declared by the presiding judge. The judge may also declare what limitations are imposed on the testimony of an expert.

The success of the designated expert in court depends upon the attention given to all facets of the case, to its details, and especially to his or her preparation. A good knowledgeable expert is, first and foremost, an excellent teacher and therefore not a member of the laboratory discussion group or of the self-serving coffee klatsch. In addition, the expert does not go into court to win or lose a case; the expert assists the attorney by presenting the truth to the court. Whether or not he or she is believable is an entirely different issue.

Although this is the first task of an attorney, the judge may prevent the expert from using too many technical terms, which may confuse the judge and the jury members; it will never impress them! Furthermore, the judge (even though opposing counsel will also perform this task) will have given great scrutiny to the qualifications of the expert, questioning if he or she is really an expert in the area of concern. Any lie or misinterpretation of the expert's past by the expert will come to the surface and could result in disqualification from the case. It is at this point that the expert may be asked if he or she knows the meaning of the word *perjury* and what part of the word he or she does not understand!

One role that the judge will fulfill, which is not often apparent to the expert, is the role of *gatekeeper*. Similarly, about eighty years ago, in *Frye v. United States*, an expert presented a novel physiological test that used the change in systolic blood pressure to detect if a witness was telling the truth or lying. The presiding judge refused to admit evidence based upon this physiological test because there was no real agreement among knowledgeable scientists on the value of its technique. In a sense, the *Frye* rule dealt with the question of which evidence can be excluded by a presiding judge (i.e., evidence not derived from normal scientific techniques). As a result, judges have the prerogative to rule out certain novel approaches or unpopular scientific opinions. Thus, the deduction a scientist or engineer makes must come from science or engineering that is sufficiently established and has gained general acceptance in its field.

Under Rule 702, Justice Blackmun intended for trial judges to act as gatekeepers. Under this rule, a judge can decide what evidence can be presented to the court and jury and, in contrast, what information must be withheld from the court and jury. Justice Blackmun made statements that were critical of the narrower *Frye* rule in the majority opinion that should make toxicologists reflect on their pending testimony. He noted, "There are important differences between the quest for truth in the courtroom and the quest for truth in the laboratory" and that "general acceptance" is not a precondition to the admissibility of scientific evidence under Federal Rules of Evidence (FRE), especially Rule 702.

The rules for experts have been somewhat modified by the Supreme Court decision relating to *Daubert v. Merrell Dow Pharmaceuticals, Inc.* (U.S. 951). In the *Daubert* majority opinion, Chief Justice William Rehnquist and Associate Justice John Paul Stevens warned that judges must become amateur scientists and (by inference) amateur engineers insofar as judges must possess the capacity to decide "whether the reasoning or methodology underlying the testimony is scientifically [and engineering, if there is such a word] valued ... and can be applied to the case." Thus, as a gatekeeper, a judge has the responsibility to determine whether or not the testimony is based on sound scientific principles and if "any and all testimony or evidence admitted is not only relevant, but reliable." This can lead to a judge's excluding unorthodox but relevant testimony. However, the *Daubert* rule does not permit a judge to decide if there is sufficient evidence for an expert to come to a conclusion.

Nevertheless, in the role of gatekeeper, the judge has the ability, if not prerogative, to make an independent assessment of the reliability of the materials used by an expert in coming to a conclusion. The judge can conduct an independent evaluation of the reasonableness of the expert's testimony and opinions. Therefore, it is also important that the scientist or engineer be fully qualified as a member in good standing of the relevant profession. Membership in a professional association is, at best, of paramount importance or, at worst, helpful.

Following from this, under Rule 702, a judge can decide not only on the qualifications of an expert witness but also on his or her methodology and scientific principles as well as the validity of the expert's conclusions. This rule permits a judge to accept one expert's opinion as being superior to that of an opposing expert. If a judge decides that any step in an expert's testimony is unreliable, that testimony becomes inadmissible.

Finally, in the role of gatekeeper, the judge can limit the number of experts who can be called to testify by either party involved in the dispute. Moreover, the trial judge must decide if the expert's scientific or engineering expertise can assist the jury while they determine the facts as related to the issues. The judge must also decide if the scientific evidence is reliable. Acceptance by the scientific or engineering community is not now the major criterion. The judge must decide if the evidence will assist the jury in deliberation.

On the other hand, the judge may actually rule out good scientific evidence if, as the judge sees it, the evidence will mislead or prejudice the jury. It is the judge's discretion to determine if novel, but credible, theories will be presented to the jury.

In summary, it is the prerogative of the judge to determine if the expert's testimony reflects scientific or engineering knowledge and whether the findings are derived from the scientific method. The judge must ensure that the proposed testimony is relevant to the task at hand and that it logically advances a material aspect of the proposing party's case. Thus, expert testimony must fit by applying accepted science and engineering to the facts of the case.

1.6 WORK PRODUCT AND ATTORNEY–CLIENT PRIVILEGE

At this time, the terms *work product* and *attorney–client privilege* need to be defined.

Once it has been determined that a particular issue is to be litigated, it becomes necessary for the expert to prepare a written report in accordance with the guidelines from the court or the relevant federal authority (Chapter 5). During this time and during the preparatory stage of the case, the expert may make initial notes and general doodling or scribbling (jottings) while gathering information about technical implications of the case. The expert records the findings, opinions, case history of similar cases, and expert views on the technical aspects of the case; all possible relevant information is collected for the purpose of the trial. These recorded documents are often referred to as a work product. Such a title may protect the jottings from opposing counsel or it may not. Opposing counsel will generally try very hard to get hold of such material on the basis that it will show if the opinion or conclusions of the expert witness were, in any way, guided or changed by the attorney for whom he or she is testifying.

Attorney–client privilege is the legal concept that protects communications between a client and his or her attorney and keeps those communications confidential. This privilege encourages open and honest communication between clients and attorneys. However, in the United States, not all state courts treat attorney communications as privileged. For instance, some states may only protect client communications; an attorney's communication will only be protected as privileged to the extent that it contains or reveals the client's communications. In contrast, other states may protect the attorney's confidential communications regardless of whether they contain, refer to, or reveal the client's communications. In addition, the U.S. Supreme Court has ruled that the privilege generally does not terminate upon the client's death.

1.7 ETHICAL CONSIDERATIONS

An expert must exhibit ethical behavior that is beyond reproach and should never give cause for disbelief in the expert's ethical considerations. For example, the goal of being an expert is to present the truth to the court—not the biased truth but, rather, the complete, unbiased, and unembellished truth. To do this, the expert must be loyal to the attorney to whom he or she reports. If contact with the expert is made by anyone other than the attorney, the expert must report this contact immediately and present the substance of the contact to the attorney.

In addition, the expert should be following the *code of ethics* of the society of which he or she is a member. Most experts are members of more than one society and manage to blend the respective codes of ethics of each society. It is good to advise the attorney of these codes of ethics when the expert is first retained on the case. An expert who disregards the code of ethics of a scientific or engineering organization can cause that organization to lose status and credibility in the eyes of the public. One unethical member of an organization can cast doubts upon the ethics of the entire membership.

A code of ethics (sometimes known as a *code of conduct*) is often a formal statement of the organization's values on certain ethical and social issues (see Tables 1.2 and 1.3). Some set out general principles about an organization's beliefs on such matters as quality, employees, or the environment. Others set out the procedures to be used in specific ethical situations, such as conflicts of interest or the

TABLE 1.2
The American Chemical Society (ACS) Code of Conduct[a]

Organization: American Chemical Society
Subject: The Chemist's Code of Conduct
Date Approved: June 3, 1994

The Chemist's Code of Conduct

The American Chemical Society expects its members to adhere to the highest ethical standards. Indeed, the federal Charter of the Society (1937) explicitly lists among its objectives "the improvement of the qualifications and usefulness of chemists through high standards of professional ethics, education and attainments...."

Chemists have professional obligations to the public, to colleagues, and to science. One expression of these obligations is embodied in "The Chemist's Creed," approved by the ACS Council in 1965. The principles of conduct enumerated below are intended to replace "The Chemist's Creed." They were prepared by the Council Committee on Professional Relations, approved by the Council (March 16, 1994), and adopted by the Board of Directors (June 3, 1994) for the guidance of society members in various professional dealings, especially those involving conflicts of interest.

Chemists Acknowledge Responsibilities To:

The Public

Chemists have a professional responsibly to serve the public interest and welfare and to further knowledge of science. Chemists should actively be concerned with the health and welfare of co-workers, consumers, and the community. Public comments on scientific matters should be made with care and precision, without unsubstantiated, exaggerated, or premature statements.

The Science of Chemistry

Chemists should seek to advance chemical science, understand the limitations of their knowledge, and respect the truth. Chemists should ensure that their scientific contributions, and those of the collaborators, are thorough, accurate, and unbiased in design, implementation, and presentation.

The Profession

Chemists should remain current with developments in their field, share ideas and information, keep accurate and complete laboratory records, maintain integrity in all conduct and publications, and give due credit to the contributions of others. Conflicts of interest and scientific misconduct, such as fabrication, falsification, and plagiarism, are incompatible with this Code.

Employees

Chemists should promote and protect the legitimate interests of their employers, perform work honestly and competently, fulfill obligations, and safeguard proprietary information.

Employers

Chemists, as employers, should treat subordinates with respect for their professionalism and concern for their well-being, and provide them with a safe, congenial working environment, fair compensation, and proper acknowledgment of their scientific contributions.

Students

Chemists should regard the tutelage of students as a trust conferred by society for the promotion of the student's learning and professional development. Each student should be treated respectfully and without exploitation.

TABLE 1.2 (CONTINUED)
The American Chemical Society (ACS) Code of Conduct[a]

Associates

Chemists should treat associates with respect, regardless of the level of their formal education, encourage them, learn with them, share ideas honestly, and give credit for their contributions.

Clients

Chemists should serve clients faithfully and incorruptibly, respect confidentiality, advise honestly, and charge fairly.

The Environment

Chemists should understand and anticipate the environmental consequences of their work. Chemists have responsibility to avoid pollution and to protect the environment.

[a] http://ethics.iit.edu/codes/coe/amer.chem.soc.coe.html

acceptance of gifts, and delineate the procedures to determine whether a violation of the code of ethics occurred and, if so, what remedies should be imposed.

The effectiveness of codes of ethics depends on the extent to which management supports them with sanctions and rewards. Violations of a private organization's code of ethics usually can subject the violator to the organization's remedies (in an employment context, this can mean termination of employment; in a membership context, this can mean expulsion). Of course, certain acts that constitute a violation of a code of ethics may also violate a law or regulation and can be punished by the appropriate governmental organ.

While following the code of ethics, the expert should observe the spirit as well as the letter of the law. If a conflict arises between the code of ethics and the law, the expert must discuss this issue with the attorney. It is far better to have any potential conflicts resolved at this point than have to answer questions related to the issues under cross-examination at the trial.

Moreover, the expert must be able to face himself or herself in the mirror every morning with no doubt that his or her dealings with colleagues and as an expert have been intellectually honest. At all times, the expert must speak the complete truth. Half-truths, partial truths, and truths that are slightly bent to please a colleague or a manager (at work) or an attorney (when retained as an expert) are not acceptable.

In terms of being retained as an expert, a scientist or engineer must avoid (1) giving false information, (2) fabricating data from a nonexistent experiment, (3) fabricating data from an incomplete experiment, (4) ignoring available data—often referred to as omitting the flyers (referring to data points that do not fit the theoretical line), (5) reaching a conclusion before all research is completed or before all data are available, (6) rendering an opinion in one case and then

TABLE 1.3
The American Institute for Chemical Engineers (AIChE) Code of Ethics[a]

Organization: American Institute for Chemical Engineers
Subject: Code of Ethics
Date Revised: January 17, 2003

Code of Ethics

Members of the American Institute of Chemical Engineers shall uphold and advance the integrity, honor and dignity of the engineering profession by: being honest and impartial and serving with fidelity their employers, their clients, and the public; striving to increase the competence and prestige of the engineering profession; and using their knowledge and skill for the enhancement of human welfare. To achieve these goals, members shall:

- Hold paramount the safety, health and welfare of the public and protect the environment in performance of their professional duties.
- Formally advise their employers or clients (and consider further disclosure, if warranted) if they perceive that a consequence of their duties will adversely affect the present or future health or safety of their colleagues or the public.
- Accept responsibility for their actions, seek and heed critical review of their work and offer objective criticism of the work of others.
- Issue statements or present information only in an objective and truthful manner.
- Act in professional matters for each employer or client as faithful agents or trustees, avoiding conflicts of interest and never breaching confidentiality.
- Treat fairly and respectfully all colleagues and co-workers, recognizing their unique contributions and capabilities.
- Perform professional services only in areas of their competence.
- Build their professional reputations on the merits of their services.
- Continue their professional development throughout their careers, and provide opportunities for the professional development of those under their supervision.
- Never tolerate harassment.
- Conduct themselves in a fair, honorable and respectful manner.

[a] http://www.aiche.org/About/Code.aspx

giving a completely opposite opinion in another, similar case, and, most of all, (7) accepting an assignment beyond his or her competence.

Ethical considerations go beyond the items listed here. For example, a question of ethics can also arise if an attorney engages a scientist or engineer as an expert to testify as a plaintiff's witness in one case and at the same time as a defense witness in another, similar case. Another question related to the expert's ethics can arise if an expert testifies in one case against an attorney but at the same time is an expert for the same attorney in a different case.

In summary, an ethical scientist or engineer does not discuss the case in which he or she is involved with anyone other than the attorney or someone the attorney authorizes. No document should be shown to anyone outside the intimate circle designated by the attorney. At times, cases are interesting and even bizarre; however, the temptation to make remarks about a case to friends should be controlled.

2 The Résumé, the Internet, and Retention of the Expert

2.1 INTRODUCTION

The first issue to be addressed by the attorney relates to qualifications and whether or not the would-be expert's qualifications are suitable for the scientist or engineer to present authoritative and believable opinions and conclusions on the case. Similarly, the scientist must know whether his or her qualifications are relevant to the issues under consideration in the case.

Like all written materials, effective résumé writing depends on organization, clarity, and how well the document is structured. For example, paragraphs can conveniently be reduced to bullet points. A scientist or engineer seeking a first appointment as an expert often attempts to fill the page using a larger than normal (twelve point) font, but font size does not compensate for quality. Above all, the résumé should highlight the scientist's or engineer's professional experience, diversity, key positions, and the manner in which the person's qualifications make him or her fit for the expert witness appointment.

To do this, the would-be expert should describe his or her job history and skill set in the best possible manner by harvesting the knowledge and experience of his or her career. Writing the résumé should use solid resources and avoid passive writing as well as wordy descriptions. The writing should be in a clear, concise manner that will truly communicate the array of the scientist's or engineer's relevant skills. Above all, the résumé should not be rejected for consideration or be placed at the bottom of the pile because it contains poor categorization, the wrong or incorrect job titles, exaggerated content, and useless information often referred to (politely) as *padding* (also called *filler*) or (impolitely) *fluff*.

The first thing that a potential employer (the attorney) looks for on a résumé is a focus and the means by which this focus will help him or her to win the case. Thus, the résumé *must* have a clear, specific statement of experience related to the case. The résumé can be tailored to match the case, as long as the experience and information presented are truthful. There is nothing wrong with having several completed résumés for various technical cases as long as the areas of expertise are within the stated purview of the would-be expert's experience. In addition, the scientist or engineer should not use a vague statement of experience in the hope that this will make him or her seem qualified as an expert witness. It will only convey to the attorney that the would-be expert is uncertain about his or her areas of scholarship relevant to the case.

It should always be remembered that the résumé will be used in court to have the judge approve the scientist or engineer and also by opposing counsel in an attempt to have the scientist or engineer disqualified from the case. At this point, it is also worth noting that it is a mistake to think of a résumé as a history of the past, as a personal statement, or as some sort of self-expression. Self-expression is all well and good in the correct context but not in the context of the expert witness.

Although most of the content of any résumé is focused on job history, the résumé submitted to the attorney must be written with the intention of creating interest and persuading the attorney that what he or she heard during the initial contact was of interest to the case. If the résumé is written with that in mind, the final product will be very different from a written catalog of job histories.

Many scientists and engineers write a résumé somewhat grudgingly and consider writing it (in the hierarchy of worldly delights) as being only slightly above filling out income tax forms. In the current context, the résumé writer is advised to muster some genuine enthusiasm for creating a real masterpiece rather than the feeble products that most scientists and engineers produce.

The résumé should establish the scientist or engineer as a professional person with high standards and excellent writing skills, based on the fact that the résumé is so well done (clear, well organized, well written, well designed, and of the highest professional grades of printing and paper).

Writing the résumé should also help the scientist or engineer to clarify his or her abilities and to decide whether or not, after the initial contact, he or she is capable of being an expert. If the scientist or engineer is uncertain (and this happens with experienced experts as well as with neophytes), further discussions with the attorney are advised. On the other hand, if the decision is negative, the scientist or engineer would do well to advise the attorney accordingly.

Assuming that the scientist or engineer reaches a positive decision and decides to accept the interview and the possible assignment, it is time to write the résumé, making sure that it is appropriate to the matter at hand. This means that the scientist or engineer should have discussed the relevant issues of the case during the initial contact. Not to do so leaves the would-be expert in the dark and unable to communicate the qualifications relevant to the case through the medium of the résumé. In addition, if the would-be expert does not determine the relevant issues of the case during the initial contact but agrees to go ahead, the attorney may wonder if he or she is retaining an expert who has no interest in the technical issues of the case but whose sole interest is the fee.

The résumé should not merely be another piece of paper that has not been reviewed or revised for several years. A résumé is a dynamic document that changes several times per year. It should not written be in a noncommittal format that makes the attorney question whether he or she is seeking to retain a professional scientist or engineer or a horse trader!

In summary, the purpose of a résumé is for the scientist or engineer to pass the attorney's screening process or to help the attorney select a suitable candidate as his or her expert. This involves (in the résumé) statements of requisite educational level and number of years' experience and giving basic facts that might influence the attorney's choice.

2.2 THE RÉSUMÉ

Providing there has been initial contact between the attorney and the scientist or engineer, the résumé will be read carefully to help the attorney decide if there is to be an interview. This means that the decision to interview a candidate is usually based on the initial contact and the overall first impression created by the résumé. The résumé must impress the attorney and convince him or her of the candidate's qualifications so that an interview is the outcome.

To write an effective résumé, the scientist or engineer has to learn how to write powerful but subtle advertising copy; as sacrilegious as this may seem, it is true. Given the fact that most scientists and engineers do not think naturally in a marketing-oriented way, they are probably not looking forward to selling anything, let alone themselves. Therefore, the scientist or engineer is advised to learn to write a spectacular (but true) résumé.

There is no need to invoke the salesman's hard sell or make any claims that are not absolutely true; however, the writer must move beyond the tendency to be modest and unwilling to toot his or her own horn. Attorneys, like people in general, more often buy the best advertised product rather than the best product. This is good news for the scientist or engineer who is willing to learn to create an excellent résumé that will be relevant to the case under dispute. With a little extra effort, the writer will usually get a better response from the attorney than scientists and engineers with better credentials but poorly written résumés.

The résumé should be a carefully edited summary of the scientist's or engineer's education and experience. In the past, résumés have been traditionally submitted on paper; however, they can also be sent in electronic form over the Internet. Whatever the form of a résumé, the idea is to select specific parts of the applicant's past that will support being retained for the case in question. A résumé presents the scientist or engineer to the attorney, who—based on his or her response to the résumé—may or may not have any further interest. It is the tool that the attorney will use most to screen would-be experts, and it is also often used as a jumping-off point for the interview.

When a scientist or engineer is preparing a résumé, he or she should keep in mind the main purpose of a résumé: to make him or her more appealing to the attorney than anyone else who is under consideration. The résumé will help showcase the relevant experience, skills, professional achievements, and educational history, as well as the potential for working on the case. In addition, the would-be expert should keep in mind that an attorney will only review a résumé for information that will distinguish the would-be expert as acceptable to the case.

Keywords that are specific to legal experience and cases should be used. Qualifications and skills should also be listed. Also, it is a good idea to list items that have been accomplished; they should not be confused with duties or tasks.

Scientists and engineers are accustomed to writing a résumé that will be read by a human resources (HR) department and by other scientists or engineers. With a little extra effort, the scientist or engineer can create a résumé that makes his or her qualifications appeal to the attorney as a superior candidate for retention as an expert. Thus, even if the scientist or engineer faces competition, with a well-written résumé, he or she should be invited to interview. This will involve learning how to think and write in a style that will be completely foreign to the professional scientist and engineer.

It is worth remembering that if a résumé is shown to three other scientists or engineers, there will be (at least) three different suggestions on how to improve it. This means that the preparer of the résumé will have to become an expert at résumé writing and make some decisions on its preparation.

2.2.1 Résumé Writing

The résumé is an advertising document—nothing more, nothing less—with one specific purpose: to move beyond the initial telephone contact with the attorney and win an interview. If it serves this purpose, it is an effective résumé.

In many instances, most initial contact between the scientist or engineer and the attorney occurs through the telephone or by e-mail because the attorney has picked the scientist or engineer through a search of the Internet. It is wise to have a résumé that confirms the attorney's initial thoughts about qualifications; to do this, the résumé must not be in the realm of fantasy but, rather, representative of the everyday life of the scientist or engineer.

A good résumé not only tells the attorney what the scientist or engineer has done but also makes the same assertion of the expertise of the writer from which the attorney will be able to determine how he or she can use that expertise. To this end, the résumé should present the qualifications of the scientist or engineer in the best possible light. The résumé should be pleasing to the eye (the author has seen résumés that resemble a piece of toilet paper—this form is not recommended) and should whet the attorney's appetite and stimulate interest in meeting the writer and learning more about his or her qualifications and demeanor.

Opinions differ, but one page is usually sufficient at the commencement of the contact, although the would-be expert is advised to explain to the attorney that he or she has a full résumé that is available on request. The information should be organized in a logical fashion, with descriptions clear and to the point. However, writing a résumé does not mean that the would-be expert should follow the rules heard through the rumor and innuendo of colleagues, acquaintances, and the like. Every résumé is a one-of-a-kind marketing communication, and it should be appropriate to the would-be expert's situation and do exactly what he or she wants it to do.

The information should be tailored to the case, there should be no fluff (i.e., indiscriminate advertising of scientific or engineering talents that may exist only marginally, if at all), and all of the information should be the truth. It is at this stage that many would-be experts do not realize that their résumés may be examined in excruciating detail by opposing counsel who will, the morning of cross-examination, have sharpened his or her teeth and be ready for the shark attack.

On the other hand, for counsel examining a résumé, he or she is well advised to determine if the package contains the truth and nothing but the truth. If the résumé contains anything but the truth, this would-be expert must be exposed and shown up for what he or she really is. The would-be expert should be made to squirm in the witness chair, and it will be at the judge's discretion whether or not he or she is dismissed and subject to charges (including charges of perjury). This may sound drastic to many readers, but the court is no place for liars, cheats, and charlatans.

Having lapsed into (hopefully) only mental and philosophical meanderings, it is now time to return to the matter at hand: the résumé. The document should be free of typographical, grammatical, and punctuation errors. The overall appearance of the résumé will affect an attorney's opinion. The résumé should be printed on high-quality inkjet or laser printers with word-processing software. Good-quality laser-printed photocopies are widely used and accepted. The preferred font is a simple, easy-to-read, eleven- or twelve-point font, such as Times New Roman.

2.2.2 Presentation of a Résumé

The presentation of a résumé is a crucial factor in impressing the attorney who may be a potential employer. In many instances, a poor résumé can mean no interview. Presentation can be thought of as packaging, but it should be honest and without question. In this sense, the résumé should present the would-be expert in the best possible light—truthful and focused on qualities that highlight the strengths needed for the case. An organized résumé will allow the attorney quickly to scope out the would-be expert's potential for contribution to the case.

Of particular interest to the attorney is the would-be expert's work history, in which jobs and experience are chronicled. This is often the most effective way of presenting the information while allowing the attorney to make an initial evaluation of the would-be expert's value to the case quickly. It will not stop opposing counsel from asking sharp and cutting questions, but it will facilitate answering such questions.

Therefore, a focused résumé may be the best type to present to the attorney. In this résumé capabilities and achievements are presented in such a way as to aim at the particular case. The purpose is to bring attention to specialized training, education, or experience related to the case. Sections of the résumé should be arranged with name, address, telephone, and e-mail address at the top, followed by relevant headings that are labeled Abilities, Achievements, Work History, and Education. Another section can be added near the end of the résumé where professional organizations may be listed.

2.2.3 Résumé Structure

When deciding on the structure of the résumé, it is often expedient to allow content to drive style, not forgetting the reason for the résumé. In other words, the scientist or engineer should first determine strengths as they are relevant to the case and then organize the résumé with headings that highlight these strengths.

Simple structures work the best. When creating the résumé, it is best to keep the statement and content simple so that the completed document is easy to read. The attorney may have limited time to make the final decision, and the best information must be provided to enable the attorney to make the correct decision.

In the first instance, the decision is not about hiring the scientist or engineer but, rather, related to an invitation to an interview. In fact, it is better to consider the résumé as a door opener. However, caution is advised because, in the last instance, the content of the résumé may allow opposing counsel to get the scientist or engineer disqualified from the case. It is unlikely that such attempts at disqualification will be made at the time of the deposition. They are more likely to be made at the trial, in which case the attorney is left high and dry without an expert witness to cover a crucial area while the opposing counsel is beaming with joy.

Information that best promotes the scientist's or engineer's qualifications to be an expert witness should be listed relative to each section. Suggestions and explanations for the various headings follow:

- Overview: a brief sentence states the main goal of the applicant, perhaps as it relates to the person's qualifications. An overview statement is optional.
- Competencies: a listing of skills, such as the ability to run certain computer programs, do desktop publishing, teach a class on a given subject, conduct research in certain areas, as well as any other valuable information, should be provided.
- Abilities: short, bulleted statements list past training, experience (certain skills, not a listing of employers), committees served on, and any leadership positions held.
- Achievements: short statements list certifications, honors, specialized training (honors classes), awards, and staffing positions of merit.
- Work history: list jobs, beginning with the most recent. Dates should be given. It is acceptable to list the years without the exact months and days.
- Education: list degrees, beginning with most recent one attained.
- Professional organizations: list profession-related affiliations and organizations in which membership is held.

Qualities and attributes should be viewed as products to be presented to the attorney. In other words, talents are assets and must be presented in the light of how they will contribute to the overall good of the company. Moreover, the résumé should be reader friendly: clear, concise, and to the point. Elaborate explanations that are time consuming to read should be avoided, as well as unnecessarily long words, sentences, and paragraphs.

The length of the typical résumé might be one or two pages. An exception to this rule is made for professions that are associated with publication of articles in journals. However, a short résumé is still appropriate, with the option for the attorney to request a longer résumé at a later date. If a longer résumé is to be submitted, a key is to have a table of contents for ease of navigation through the document. Examples are résumés of authors who list books, musicians who list published song titles, and scientists who list published research.

Goals and objectives can be tailored to fit the case under consideration, although achievements will speak for themselves without embellishments. In addition, a list of references should be prepared for the attorney if he or she requests them; however, many résumés do not list references. If they are provided, the attorneys used as references should be made aware that they may be contacted ahead of time. Above all, it is important that the résumé be professional, because it is the first sample of the applicant's work that an attorney—a potential employer—will likely see.

2.2.4 Education

In many résumés, the education section should precede the experience section. However, assuming that the would-be expert has a strong work history and related experience, it is advisable to place the experience section before the education section. That being the case, the education section still requires organization. A recommended layout is to (1) begin with the name and location of the institution attended, listing the ones from which a degree was obtained, and (2) give the degree and the month and year in which the degree was earned. The name of the degree may be written out in full or abbreviated; consistency of style is the key. Degrees are often, but not always, listed in reverse chronological order.

2.2.5 Experience

This section often starts with the most recent job first. Each employer should be listed, along with the city and state, job title, and dates of employment, as well as details of accomplishments and any special skills used. Experience is assumed to be full-time paid work experience unless stated otherwise. Experience may be described in paragraph form or in bullet points; the latter style is recommended for ease of reading.

In the various job descriptions, the would-be expert should mention skills that directly relate to the present case. If previous jobs were not directly related to the case, it is permissible to emphasize skills used in previous jobs that could be used in the case (transferable skills). If there is no related previous experience, the would-be expert may be well advised to walk away after informing the attorney that his or her skills are not part of those required for the case. If the would-be expert does not do this and the attorney unknowingly hires him or her as an expert, the expert can be sure that opposing counsel will raise these issues during cross-examination, which could well present the embarrassing situation of being excused from duty for being unqualified for that case.

2.2.6 Accomplishments

Terms such as *responsible for* or *duties included* should be avoided. Instead, when duties and accomplishments are described, the description should begin with an action verb, followed by what was actually done, followed by the result. The attorney who is considering the scientist or engineer for an expert appointment needs to know what the prospective expert did above and beyond any minimum requirements. This section should include problems solved, special projects, special assignments, training, and travel—anything that makes the would-be expert special compared to all other candidates. The attorney is not interested in vague or blustering windbag statements that lead nowhere and, when analyzed, mean nothing.

2.2.7 Professional Organizations

Most scientists and engineers belong to a professional society that requires presenting credentials to a membership committee. However, the expert must be able to explain to the judge and the jury the difference between societies where membership is available for a fee and what they mean to the scientist or engineer (annual meetings and newsletters and magazines that keep the members up to date in recent developments of science and engineering) and societies that present members with a professional qualification that is subject to examination on a specified basis.

2.2.8 Honors and Awards

Commendations, awards, and honors or any formal recognition that supports the objective of submitting the résumé should be mentioned. These could appear in a separate section or can be placed into the work experience, education, or activities sections.

2.2.9 Résumé Ethics

Consider the old fable: *Put anything you wish into your résumé because there is no way anyone will find out.* There is nothing further from the truth, and the would-be expert who has done this must live with himself or herself and the ignominy of being found out in public for lying on a résumé. As tempting as it might be to insert little white lies on a résumé (the advertising world often refers to such as *fluff*), such as overstating knowledge of required technical matters, a certification, or extending dates at a former employer, this is not to be condoned.

The would-be expert who lies on a résumé has to discuss his or her résumé under oath in the witness chair. The risk of being discovered is real, and the odds of getting away with listing false information on a résumé are serious. Rather than go any further, this type of person should acquaint himself or herself with the definition of perjury and the consequences of being found guilty of this crime.

There is only one reason for opposing counsel to conduct a background check of an expert witness: to discredit the witness and win the case. Perhaps this is two reasons, but in reality the outcome is the same.

Some attorneys have never experienced a dishonest expert who incorporated falsehoods into his or her résumé and do not routinely verify work histories and the validity of credentials. This can be a costly problem, and it would be within the attorney's right to seek compensation for money spent on such a witness.

As would-be experts construct a résumé and its contents, there are ways to verify all the information, words, data, and facts. For example, a yes answer to each of the following questions means an honest résumé has been completed:

Is it the truth? (This is also the first test of the four-way test for members of Rotary International.)

Is this a document of which the writer can be proud?

Is the would-be expert willing to have the attorney use the information to evaluate him or her for a position?

Is the would-be expert willing to define or defend the résumé during an interview?

Can the résumé be validated through a letter of reference; citations to published work; or a telephone call from a professor, supervisor, employer, or someone else who knows the would-be expert?

In summary, the résumé should always present an honest and accurate picture of education and accomplishments. The content of the résumé must be supported by any possible documentation. The expert must be able to define, defend, or describe in detail all that is in the résumé.

The moral of this section is that an expert should not risk his or her future by lying about the past. Honesty is always the best policy. The consequences of false or misleading résumé claims can be serious.

2.3 USING THE INTERNET

For the scientist or engineer who wishes to be known as having the willingness to appear as an expert witness, the *Internet* (or the *World Wide Web*; the terms are used interchangeably) can be a gold mine for soliciting contact from attorneys. There are few professionals in any discipline who do not use the Internet these days. Many attorneys browse through the available information (*surf the Net*) for potential expert witnesses.

Advertising on the Internet is a recent way for the scientist or engineer to advertise and tout his or her abilities. However, because of the drastic differences in popularity between sites, only the top, narrow percentile of Web sites may generate sufficient interest for serious contact from potential clients.

Advertising through a personal Web page or through an organization that specializes in expert witnesses can pay dividends for the would-be expert. The benefit of Web advertising is that the number of site visitors can be tracked. However, Internet advertising should be valued in terms of the business it creates from the new users it attracts to the site. This value serves as a guide to the efficiency of the Web site. Many Web sites generate only a small number of visitors, while others generate thousands of visits. In the former case, the site owner needs to evaluate the key words (or the construction of the site) and deduce why the site is not contacted frequently. Even though changes are made to an infrequently contacted site, the site owner should always be willing to change the site structure (often on an annual basis) to maintain pace with the changing world of the expert witness. Web advertising can work very well and, although not always successful, can be one of the largest contributors to the scientist's or engineer's efforts to garner clients.

The Internet differs from other forms of advertising because it is mainly a cognitive medium rather than a verbal (word-of-mouth) or an emotional medium, such as television. Scientists and engineers are unlikely to advertise on television, so this form of advertising can, in the present context, be dismissed. However, attorneys use television as an advertising medium, and such an advertisement (which always has a telephone number, usually an 800 [toll-free] number) might be the option for a potential expert witness to make the initial contact.

However, users need to recognize that the Web is a user-driven experience insofar as the user is usually on the Web for a purpose. The user is on the Web to get something done and is not likely to be distracted from the goal by an advertisement. In addition, the user is also *actively considering where to go* next, and this active user engagement makes the Web more cognitive.

There are many ways in which the Web can be used fruitfully in marketing. Most important, in the current context, are Web sites where the entire site is devoted to promoting an organization's expertise in providing expert witnesses for various cases where specific disciplines are required. Such sites focus on customer support and service, including detailed qualifications of their (unlisted) experts and supporting information for the attorney to consider.

Although many scientists and engineers have focused on the use of the Internet for advertising consulting services, others (with a view to being retained as an expert witness) have learned that the most innovative use of the Internet is through eye-catching (or mouse-click-catching) keywords. However, the two keywords that are often omitted are *expert witness*. Attorneys may not be sufficiently skilled in scientific and engineering terminology to differentiate between their needs and the words used on a Web page. The words *expert witness* will always ring a bell for an attorney with those requirements. Whether or not the scientist or engineer has appeared as an expert in the necessary discipline, he or she may be able to point the attorney to a colleague or certainly in the right direction.

In the past decade, however, thousands of scientists and engineers have created Web pages and many more individuals will create Web pages in the nest twelve months. Any scientist or engineer who has not thought seriously about implementing a Web page to tout his or her services as an expert witness should certainly consider this option.

Individual attorneys and law firms have been attracted by the demographics of the millions of users of the World Wide Web who tend to be more mature, more educated, and more affluent than the population at large. Internet-savvy attorneys, scientists, and engineers have learned that Web pages can be excellent ways to deliver and enhance services. A Web page offers the scientist or engineer the chance to create a presence in and gain access to this expert witness enterprise zone, and it offers the attorney the chance to find a scientist or engineer willing to serve as an expert.

2.3.1 The Web Page

The Web page serves the scientist or engineer as a means of advertising his or her expertise in terms of the types of work performed. There are two types of Web pages: (1) the *static* Web page, which is basically an electronic business card or brochure; and (2) the *dynamic* Web page, which adds a number of enhancements and interactive features.

A static Web page generally is a simple conversion of a marketing brochure into electronic form. Such a page might contain biographical information about the scientist or engineer, areas of expertise, contact information, and other general marketing information. In some cases, a static page is no more than an electronic business card or billboard. Other static pages are electronic copies of existing marketing brochures or recycling of existing marketing materials. The benefit of a static Web page is that it is easy to read and less flashy and gaudy. The drawback to a static Web page is that the language may be too technical and there may be no reason for the expert-hunting attorney to visit the page more than once.

A dynamic Web page includes much of the same material found in a static page but adds a number of important enhancements, such as articles written by the site owner, e-mail newsletters, lists of links to other Web pages, and other information helpful to the site visitor. Each of these enhancements is designed to prompt some form of interactivity or to encourage an attorney to contact the site owner.

A scientist or engineer can give attorneys a reason to return to the Web page by providing interesting and changing content that highlights the specific aspects of expertise. The Web site should also provide easy and speedy navigation through all of the pages constituting the site so that visitors (i.e., attorneys) can easily find content useful to them and e-mail contacts to request more information or follow-up with the scientist or engineer.

Many scientists and engineers have generated clients from their Web pages; the jury is still out (pun intended) as to the value of a static Web page versus a dynamic Web page. In spite of the advantages and disadvantages presented for these sites, both have merit. In fact, the numbers of visitors to the site do not always tell the true story. Over a given period, one site may have several thousand visitors that have resulted in one appointment as an expert witness. Over the same period, another site may have had only hundreds of site visitors but resulted in several appointments as an expert witness.

In summary, there are many innovative and useful Web pages belonging to scientists and engineers. Visiting other discipline-related pages will give the would-be expert an idea of some of the potential benefits of a Web page.

2.3.2 The Specialty Page

Another type of Web page is the *specialty* Web page. This is devoted to one specialty topic; for example, instead of the more general topic of reservoir engineering, a specialty page may describe the engineer's expertise in a particular type of oil recovery process. A specialty page can be a stand-alone page done by an individual scientist or engineer, or it can be a subpage on a more general Web site.

Specialty pages tend to be built around a set of articles written by the scientist or engineer or a set of links to another Web page owned and operated by the individual scientist or engineer that deals with the same specialty topic. Specialty pages generally grow out of the scientist's or engineer's professional interest or as an attempt to fill a perceived need for such expertise as described on the page.

Over time, a specialty page might grow to include related links, other articles, and professional pointers that also lead to other consulting appointments rather than the expert witness appointment alone. A well thought out specialty page can be an excellent starting point that allows the attorney to read a quick overview of a specialty area and that can also be helpful to attorneys seeking technical and more than the usual in-depth treatments of a specialty subject relevant to a dispute.

A Web-based specialty page is typically designed to help attorneys through the inclusion of educational materials and published material, which all enhance a scientist's or engineer's reputation as an expert in a specialty area. Many specialty pages are not designed with the primary intent to produce immediate retention as an expert by an interested attorney. The initial telephone contact is still a "must," after which an attorney will have a good idea that the would-be expert is worth the risk of the attorney's expenditure on an interview. Furthermore, by developing a reputation though a specialty page, a scientist or engineer might expect to get referrals or new clients as an indirect benefit, in much the same way as would a scientist or engineer who lectures at law-related seminars.

However, not every specialty area of science and engineering has a Web page devoted to it. With a little effort and persistence, the scientist or engineer can develop a focused and eye-catching (mouse-click-catching) specialty page. Every scientist and engineer has a unique perspective and expertise and is capable of producing a creative and interesting Web page.

2.3.3 Starting a Web Page

The best way to start a Web page is not to concentrate so much on the mechanics of creating the page but, rather, to spend some time on the Web looking at as many scientific and engineering-related Web pages as can be found. In fact, a number of Web pages will direct the curious scientist or engineer to some of the best technology-related Web pages. After researching a number of pages, the would-be expert will know if having his or her own Web page is going to reach out to attorneys searching for experts. In addition, the scientist or engineer will gain a good understanding of what seems to work for other scientists and engineers as well as those Web page features that might work well on the page.

The first item of decision for the Web page is to make sure of the goals. A scientist or engineer who is hesitant about becoming an expert witness should take the time to come to a strong understanding of what is involved. The would-be expert should make a rough hand-drawn sketch of the page using graphics, layout, and possible navigation to other sites. During this time, the page owner thinks at all times like a nontechnical user (e.g., an attorney): is the page easy to navigate? Is the most valuable information easy to find? Will it be easy to contact the scientist or engineer from the information provided?

Once the basic content and format are decided, the next item for consideration, before the page is implemented, is whether or not to self-create the page or to use an expert. Creating a Web page is now much easier than in the past, and the programming language for creating Web pages (HTML [hypertext markup language]) is not difficult to learn. New Web page creation programs, such as those obtained from Microsoft and AOL, to mention only two, make the process even easier because they eliminate the need to learn the underlying HTML code before creating the Web page.

On the other hand, a Web page consultant might produce a Web page with graphic design and features that the scientist or engineer could not produce. A professional designer might also help avoid problems that would only be discovered through experience. The unseen benefit of using a professional is that he or she will most likely offer advice on the readability of the page and help focus the page on what the scientist or engineer really needs.

A second, extremely important issue relates to keeping the page updated. As court cases and the need for experts evolve, the would-be expert should have a page that regularly provides new or updated content. Even if the page was created by a professional, the scientist or engineer will need to decide if that person will update the page on an ongoing basis or whether it is more reasonable to perform an update himself or herself. If updates are not done regularly, many attorneys may not return after finding that a page has not been updated for a year or more.

Finally, getting started on a Web page can be a happy or a traumatic experience, and several tips are worth noting for the would-be page owner. The scientist or engineer should not be overwhelmed by trying to design a perfect, feature-laden Web page that is never actually complete. The page should be initiated and then periodically improved as the owner finds out the approach that works and that which does not work. There is a need for a reliable host for the Web page. Most Web pages are hosted by a third-party provider. There are also ethical issues. The scientist or engineer will need to know the applicable rules and monitor any changes or developments of the rules.

2.3.4 Getting Publicity for the Web Page

The best Web page ever designed is no good if it cannot be found and no one sees it. The scientist or engineer will need to make sure that the people for whom the page is designed (attorneys) can find it. The guiding issue is to know the target audience and how that audience will find the page and read the information contained there. In fact, planning for getting publicity for the page may be more important than planning for the actual design of the page.

Once the Web page is online, an announcement about the page is necessary. The address of the Web page should be put on business cards, stationery, brochures, and any other advertisement. If the would-be expert creates a newsletter, he or she will want to do a feature article on the new Web page as well as create an e-mail version of that newsletter that can be subscribed to from the Web page. The would-be expert will do well to remember that the Web page is not only an adjunct to a business card or brochure but also an extension that can be reviewed quickly and conveniently by attorneys or clients who prefer Web surfing as the means of locating a potential expert.

Furthermore, it is vital to inform existing clients about the page. One of the best uses of a Web page is to help existing clients find out about other services the scientist or engineer can provide. The best way to increase business is to resell to existing clients by providing other services to them that they did not know could be provided.

For new clients, a quandary exists. There is no single index to the World Wide Web. Attorneys find information using search engines such as Google. Although many search engines are available, the Google search engine is the author's own personal preference, and mention of it here should not be used as an endorsement of this particular search engine. One often hears the expression "I'll Google you on the Web," which is not meant to be personal but, rather, a sign of curiosity about the qualifications of the scientist or engineer. Search engines allow the searcher (the attorney) to use keywords to search a database consisting of indexed information from actual Web pages and to produce a list of links to Web pages that best match the search request.

The scientist or engineer should also consider getting his or her Web page listed in one of the directories of expert-related Web pages. All that is needed is to contact those Web pages and ask if the page owner will add a link to the expert's page. This is, in fact, advertising, and an annual fee will most likely be charged.

An attorney searching the Web for a prospective expert will use one or more keywords. Assuming that the scientist or engineer wishes his or her page to be found, he or she must learn enough about search engines to find out how to guarantee that the Web site will found by someone doing a search using those keywords. This may require frequent reconfiguration of the Web page if no contacts have been made.

On the other hand, if the scientist or engineer wishes the Web page to be found because it contains specialty information, the page must be designed so that attorneys looking for that type of information will find a link to the page when they use the most likely keywords on one of the major search engines. It is essential for the page owner to remain focused on how a searcher may find the page.

The scientist or engineer will get the best results not only by developing and executing a plan to gain publicity for the site but also by monitoring the results. There is a need to be innovative and to remember that getting publicity for the Web page is an ongoing process.

The following points are offered as a guide to good Web design, in order of importance: (1) the page must be found; (2) download time must be rapid; (3) presentation must be succinct, with not too many pictures and using, but not overdoing, graphics; (4) navigation must be easy; (5) the page must be informative and demonstrate that the page owner is a specialist; and (6) full contact address, phone/fax numbers, contact name, and e-mail address or message box must be provided. The page address and date must also be inserted in any printout of the page.

The use of the Internet is dramatically changing the ways in which all legal business is conducted and the means by which scientists and engineers with the relevant case-related qualifications are located and retained as experts. It is impossible for the Internet not to have a similarly dramatic impact on the business of being an expert witness. The Internet is certainly an advertising vehicle that should be used by all scientists and engineers who have a yearning to advertise themselves as consultants.

A little more than a decade ago, ownership of a Web page was, perhaps, a novelty. Then it became a trend. Now it is a necessity.

2.4 SELECTING AND RETAINING AN EXPERT

Having a relevant résumé available for perusal by an attorney and having constructed an eye-catching (mouse-click-catching) Web page, the scientist or engineer may have made the initial contact and been invited to an interview. By this time, the attorney has developed a list of scientific and engineering prospects. He or she may have talked with other attorneys, evaluated the prospects, and decided that one or two of the prospects will be interviewed. The attorney will also need to be assured that the would-be expert is not economically, socially, or professionally involved with anyone involved in the dispute.

Prior to the interview, references and credentials of each candidate should be checked and information contained in the various *curricula vitae* should be checked thoroughly. Published articles and a description of previous work (litigation or otherwise) in the area of expertise should also be investigated; references should also be checked. This information will assist the attorney in deciding which of the candidates should be retained and whether the retention should be as a nontestifying expert, a consultant, or a testifying expert for the trial.

Any irregularities that have been included in any of the documents supplied by the expert for the attorney's perusal should be investigated fully. Alternatively, these irregularities can and should result in summary dismissal of the candidate from any further consideration—hence (scientists and engineers, please note), the need for a résumé *that is as clean as the proverbial whistle.*

2.4.1 The Interview

Once the list of potential candidates has been narrowed to a manageable number, each candidate should be personally interviewed by the attorney. Before the interview, the attorney may send a carefully selected packet of introductory materials to each candidate. Because all of the material reviewed by the expert may be subject to discovery (more on that later), only materials from the public record should be sent to the candidate. In addition, the candidate is well advised not to make any notes on these introductory materials, because any notes may be discoverable if they are pertinent to the case.

During the course of setting up the interview, the candidate and the attorney will learn several things about each other. For example, if it was difficult to come to a meeting of the minds about an interview date, it may be difficult for the one to have access to the other as the case evolves. On the other hand, if the candidate reviewed and assimilated all of the information that was sent by the attorney, it can be reasonably concluded that the candidate is diligent.

At the beginning of the interview, the candidate and attorney should agree that all matters discussed will be treated as confidential. The candidate should not take notes, thereby removing the potential for embarrassing questions by opposing counsel at depositions and at trial. The best approach is for the candidate to maintain an open mind and not form any opinions until he or she has been retained, completed the investigation and analysis, and formed conclusions.

During the interview, the existence of conflicts should be explored and discussed. It is preferable that the candidate not have any real or potential conflicts. Such conflicts can arise if the candidate has had social or professional dealings with the opposing disputant or with opposing counsel. In addition, recognized experts may be contacted by attorneys representing both sides of a dispute, and any such contacts should be revealed.

During the interview, the attorney may ask the candidate if, based on the materials received and reviewed, he or she has formed a preliminary opinion about the dispute. If the preliminary opinion is tentative but favorable to the attorney's client, the candidate may be asked what additional facts might strengthen his or her opinion. On the other hand, if that preliminary opinion is unfavorable to the attorney's client, termination of the interview is appropriate. However, it is preferable and a matter of professional courtesy that an adverse opinion be made known by telephone to the attorney prior to the interview.

There is a very high value to the interview process. The candidate and the attorney have the opportunity to evaluate each other (the attorney evaluates individual candidates and is able to compare one to the others). The interview also provides the candidate and the attorney with the opportunity for confidential discussions of the issues that have arisen from the dispute.

2.4.2 After the Interview

As a result of the interview, the attorney will have finalized his or her decision about the candidate. Assuming that hiring or retention of the scientist or engineer is the decision coming out of the interview, the attorney may delay the decision as to whether the expert will testify until fairly late in the discovery process. In the meantime, the attorney may employ the scientist or engineer as an expert or as a consultant from an early time in the life of the litigation, even before the complaint is filed.

In this context, the expert can assist the attorney in the technical aspects of the case and whether or not the expert believes the case to be technically prosecutable (or technically defensible), as well as whether technical aspects can be identified as justifiable. In some cases—particularly highly technical ones—when the consultant is presenting advice to the attorney that is outside the expert's line of expertise, the filing of a complaint based on such advice without consultation with an appropriate expert can damage the credibility of the case.

Before this can happen, the scientist/engineer and the attorney should determine the suitability of the consultant expert for the case and whether or not he or she is indeed qualified to testify and will be accepted by the court as an expert. Assuming that the consultant is qualified and whether or not he or she is going to testify, one of his or her functions is to teach the attorney about the subject matter of the litigation so that the attorney can persuasively cross-examine the expert retained by the opposing counsel.

Even though the attorney has mentally decided the expert for the case, he or she may still be hesitant (with some justification) to use the scientist or engineer to his or her maximum potential until the attorney becomes more familiar with the expert's qualifications. In fact, the attorney may decide that the retention of two (or more) experts for the same issue is appropriate. One expert can serve the function of a consultant and the other that of the testifying expert. Although none of the initially retained experts may ever testify in the case, it may be wise to decide at the beginning of the case that one of the candidates will serve as a consultant whose opinions and work product will not be discovered by opposing counsel.

The consultant can perform a valuable role for the attorney. He or she can assist in the formulation of a technical complaint or recognize defenses to such a dispute. When he or she is involved early in the litigation, the consultant becomes familiar with the facts as they are developed; this can better prepare the expert to testify and to point out new (or fruitless) lines of investigation. The consultant may also be called upon to assist the attorney in developing a written discovery plan as well as assist with drafting interrogatories and requests for the production of documents—all of which serve to assist the expert to form an opinion.

The consultant can also assist the attorney in preparations for the deposition of expert witnesses retained by the opposing counsel. With regard to factual witnesses, the consultant should be asked what questions he or she would like the attorney to ask the fact witness. It goes without saying that the consultant can be extremely helpful in assisting the attorney in taking the opposing expert's deposition. Indeed, assuming that his or her expert will testify, the attorney may want to have the consultant attend the deposition. The consultant not only can suggest questions but also can assist in understanding the words and phrases testified to by the deposing expert and in following up on his or her responses.

During the course of the litigation, the consultant can provide useful assistance in possible settlement negotiations. Experts, particularly economic experts, can assist in valuing the potential claims so that the attorney can advise the client as to what an appropriate settlement posture might be. The consultant can also explore alternative theories and hypotheses and perform research projects. If the results of that research or opinions formulated by the expert are unfavorable, then the attorney can abandon those theories with some confidence that the results of such work will not be discovered.

Ordinarily, counsel will attempt to ensure that a non-testifying consultant's work is not subject to discovery. To accomplish this, counsel should insulate the consultant from the trial expert. Thus, the result of the consultant's work should not be reported to the trial expert or possibly even to the client. Ideally, the trial expert would not even be told about the existence of the consultant.

As part of this expert–attorney interaction, it is preferable for the attorney to involve the client at every step of the selection and retention process. It may even be advisable for the attorney to introduce the expert to the client. This may not be preferable if the client has been involved in a serious accident such as a fire and has been burned badly. Even though the issues are related to the flammability of the chemicals that caused the fire, opposing counsel may claim that the sight of the client in a hospital bed swayed the expert emotionally and thus such testimony should be disqualified on the basis that it is not based on chemical facts but on emotion.

2.5 RETENTION OF THE EXPERT

In approaching the retention of the expert selected to testify, counsel must assume that all materials provided to the expert; every document generated by the expert; all materials circulated among the expert, counsel, and the client; every item in the expert's file; and the content of every conversation among the expert, counsel, and the client or anyone else concerning the expert's work in the case will be discovered. Therefore, from the inception of the relationship between counsel and the expert, care must be taken to minimize the generation of written material that could be used to impeach the expert.

2.5.1 The Retention Letter

In most, if not all, cases, the consultant expert should be retained by the attorney rather than by the client. This will give the expert better control of the relationship.

Similarly, retention of the consultant by counsel will better ensure that the work product doctrine will attach to the work performed by the consultant.

The first rule regarding retention letters is that counsel should assume that the letter will find its way into the hands of opposing counsel. Therefore, the retention letter should be drafted with care. The letter should be short and to the point and contain only the fundamental points of employment. These points should include the expert's fee structure, the rate of billing, the manner of billing, and initial instruction on what the attorney wishes the expert to do. The letter should commit the expert to treating all aspects of the engagement as confidential.

Omitting references to any facts or materials provided to the expert in the retention letter means that letter will not provide grist for the cross-examination mill. On the other hand, the expert should be aware that this short letter provides little or no guidance to the expert as to what work he or she is expected to do and the budgets established for that work. Such matters can be worked out in person with the expert, either verbally or using a blackboard.

2.5.2 Document Control

Counsel should maintain an inventory of all materials provided to the expert in connection with the case. Thus, if or when the expert is deposed, both counsel and the expert will know which documents need to be produced for opposing counsel.

The expert should be provided with all relevant documents or facts that the opposing counsel has or will obtain during the course of the litigation. Such documents and facts should consist of not only the helpful ones but also the harmful ones. In this way, the expert will be able to provide counsel with ideas on how to rebut harmful facts and will be better able to respond in deposition or trial to hypothetical questions or statements containing unhelpful facts. Once again, counsel should always assume that all documents given to the expert are discoverable unless counsel has elected to have that expert serve only as a consultant.

The expert should be advised that no reports or studies should be prepared without the prior approval of the designated attorney and that any work product so prepared should be factual and not express any opinions or conclusions. If opinions or conclusions are expressed in writing, all of the backup data should be clearly identical. All materials prepared by the nontestifying expert should be labeled "Privileged and Confidential—Prepared at the Request of Counsel." For this purpose, the use of a rubber stamp may be appropriate.

The expert should be instructed to exercise extreme caution in any interim views he or she forms and any notes that would ordinarily be prepared and kept in his or her work. Casual comments that have not been thought out can, in the hands of a skilled opponent, provide sufficient impeachment materials to destroy the effectiveness of the expert.

The expert should be asked for the names of the people who will be assisting him or her on various projects. Counsel should insist upon meeting with all of the expert's assistants who will be working on the case so that they can be instructed on issues of confidentiality, document control, and document generation.

2.6 CONCLUSIONS

The selection and retention of experts is a serious matter requiring a great deal of thought and work. Following the steps suggested in this chapter can make the task of finding the right legal–technical match somewhat easier for the expert and for the attorney.

Opposing counsel has a much broader right to obtain discovery from a testifying trial expert than from a consultant expert. Rule 26(b)(4)(A)(i) of the Federal Rules of Civil Procedure (FRCP) allows discovery of the subject matter of the expected testimony of a trial expert, the substance of the facts and opinions to which the expert may testify, and the grounds for each opinion. On the other hand, discovery of a nontestifying expert (hereinafter sometimes referred to as *consultant*) is allowed only if the expert made a medical examination of a party or "upon a showing of exceptional circumstances under which it is impracticable for the party seeking discovery to obtain facts or opinions on the same subject by other means" (Federal Rules of Civil Procedure 26(b)(4)(B)).

If the attorney has decided that the expert will not testify and is reasonably certain that he or she can avoid problems, such caution may be unnecessary. However, even in that event, the consultant should be instructed as to the importance of confidentiality and specifically told not to discuss his or her work on the case with anyone.

3 The Expert Witness

3.1 INTRODUCTION

An expert witness is often difficult to define or describe. Some observers may contend that an expert is the person from fifty miles away who comes around occasionally to voice the same opinions he or she has been voicing all along, just to help support his or her opinion and get paid. Others argue that the idea of an expert witness is questionable because that individual could be viewed simply as a hired technical mouthpiece for either the prosecution or defense. Yet other attorneys and courts insist that using an expert witness is the only way to take a technical case beyond the realm of theory and legal jargon and make it applicable to those hearing the arguments and assigned the task of passing judgment.

In California, the courts have defined an expert as someone who has education, training, or experience in a particular subject that the average person does not have. The expert can be from any profession or of a special skill as long as he or she has the ability to assess and evaluate conditions on a case related to his or her individual field of specialty.

However, as noted in Chapter 1 and for the purposes of this text, an expert witnesses is a scientist or engineer who has been designated by the court (the judge) as acceptable for the case and who will testify under oath about the technical details of the case and state his or her opinions and conclusions about the matters under question. This chapter focuses on the scientist or engineer who has been retained to be a designated expert witness.

Furthermore, the scientist or engineer who accepts the designation of *expert witness* must be willing to accept such opinions as those enunciated in the first paragraph. Moreover, he or she *must* have sufficient knowledge of the facts of a case to evaluate them, and this knowledge permits forming opinions and drawing conclusions.

However, the scientist or engineer must remember that experts are not brought into a courtroom to argue the facts of a case. They do not present evidence as do other witnesses or police in the situation involving a criminal investigation. The expert is present to clarify the technology being used and the way the scene was measured or examined. In fact, the scientist or engineer will do well to remember that he or she has been retained to appear as an expert as an audio and visual aide to help the judge and jury understand the technical aspects of a case.

In most technical cases, both the prosecution and defense bring in scientific or engineering experts to support their respective sides of the case. As professionals, scientists and engineers are expected to be loyal to the knowledge they bring to the case rather than to either side in the courtroom. The scientist or engineer as an expert is present to help the judge and jury obtain an accurate assessment of the technical issues of a case. The finder of fact is the jury—or the judge, if it is a bench trial. After presentation of the opinions and conclusions, it is up to the trier of fact to determine if it has been helpful and assists in analyzing the facts being offered.

The issues of what qualifies a scientist or engineer to be an expert may be as subjective as the case itself. Although it is common to bring in academics to explain situations ranging from scientific and engineering theories (which may help or hurt the case because judges and juries can become confused by academic pontificating without ever reaching the point), many attorneys contend that the best experts are the people who work in the various industrial activities that bear a strong technical relationship to the technical issues of the case. For example, a scientist or engineer with working experience in the petroleum industry would be preferred over an academic who has never been out of the classroom or ever visited a refinery. However, the academic witness might, in turn, be better suited as an expert in such a case than a corporate executive because of a technical understanding of and knowledge about the issue in question.

There are real experts and quasi-knowledgeable experts, and there are well-intentioned scientists and engineers who are not qualified in a particular field but who can talk a lot. The latter are soon shown for what they are in court: ineffective and damaging to a case. Also available are those scientists and engineers of a less ethical persuasion who, for a fee or because they firmly believe that they can tell any story in court, are willing to testify and to present their opinions. Therefore, the qualifications of the expert become critical to enabling the jury to understand and believe his or her scientific testimony.

Finally, there are the scientists and engineers who, having worked for a company for thirty or more years, want to find something to do after retirement. These unfortunates can find that the romance of being engaged as a nontestifying expert or testifying expert witness (Chapter 1) fades when they have to struggle with a mental conflict that requires differentiation between what they can say and what remains company confidential. Such experts are, of necessity, short lived.

It is the business of the attorney to determine which of the preceding experts he or she can engage to help with his or her case. If the attorney is satisfied with the scientist's or engineer's credentials, he or she must be able to present the expert and those credentials in terms of years of experience and years of study; this must assure the judge and jury of the would-be expert's credibility.

However, taking into account any challenges by the opposing counsel, the judge will make a determination whether or not the scientist or engineer is an expert suitable for the case. During his or her determinations, the judge will be looking for experience and credibility and will thoroughly consider qualifications and experience that support the scientist's or engineer's designation as an expert.

3.2 QUALIFICATIONS AND EXPERIENCE

Expert witnesses play a vital role in the judicial system, and this is especially true for the scientist or engineer who opts to be an expert witness. They can prevent the filing of unmeritorious lawsuits and even change the course of a case. For one part or the other, engaging the expert whose credentials are most relevant to the case is a major issue. Expert witnesses can often make or break a case when it comes to prosecuting or defending a case in a court of law. Now more than ever, issues in litigation are requiring the services of expert witnesses to enlighten judges and juries on technical matters and standards of care for industry issues related to the cases brought before them.

Most critical to choosing an expert is the ability of the scientist or engineer to assimilate the facts, abstract the relevant issues, synthesize facts, reach a conclusion, and present opinions and conclusions in a manner that in understandable to the court (the judge and the jury). Openness, honesty, forthrightness, and a genuine desire to help a jury or judge decide a case are valuable traits that cannot and should not ever be overlooked or underestimated.

When choosing an expert witness, attorneys look for those with qualifications that will help them establish who or what might be involved or the cause of issues in the lawsuit, thus adding substance to their position or allegations. In some cases, the scientist's or engineer's level of education (i.e., BS, BSc, or B.Eng. vs. PhD) is far less important than the scientist's or engineer's knowledge and level of experience in the necessary field or discipline.

3.2.1 Qualifications

The first issue to be addressed by the attorney and the scientist or engineer relates to qualifications and whether or not the would-be expert's qualifications are suitable for the scientist or engineer to present authoritative and believable opinions and conclusions on the case.

The attorney should first determine if the would-be expert can present a minimum of *education* in terms of degrees awarded and the institution where they were awarded. Similarly, the attorney must be able to present this information to the court and have the scientist or engineer accepted as an expert on the case.

The major subject of study, or *area of scholarship,* should be related to the issues of the case. A passing interest or a hobby study is usually of little consequence and relevance in qualifying a scientist or engineer to be an expert. For example, an interest in figure skating and even being a figure skating judge does not give a scientist or engineer the necessary qualifications to testify on a case related to injuries sustained during figure skating.

The potential expert should assess the qualifications that give him or her a right to place his or her experience and credibility forward in order to be engaged by an attorney to have an opinion on a case. He or she must be able to present to the court and to a jury a minimum of acceptable qualifications that is required to present a credible opinion. The first aspect of qualifications pertains to education. The jury will assess an expert based on the subject matter of the expert's degrees and the university that awarded the degrees. Information such as the year that the degrees were awarded, the major subject, and the minor subjects is important.

Advanced degrees in environmental science and environmental engineering are, generally speaking, relatively recent phenomena, commencing in the 1970s. If the expert is the recipient of such a degree, members of the jury need to know that the expert has a good basic grounding in a scientific or engineering discipline. There was a time (not so long ago) when the recipients of degrees that dubbed the recipient *environmental scientist* or *environmental engineer* were noted to be weak in basic scientific and engineering disciplines. Fortunately, most universities have taken steps to rectify these apparent weaknesses.

The judge and jurors will be interested in *honors* conferred upon the expert. These should be mentioned in a résumé and not overplayed. Judges and juries may be suspicious of too many honors. The scientist or engineer who has several honors to his or her name may be, in the minds of a judge and jurors, more credible than a scientist or engineer who lists several pages of honors.

It is not a good idea for the scientist or engineer to attempt to show that his or her breadth of knowledge blankets the world. Juries frown on experts who cover too many diverse topics and appear to be global experts. A good rule of thumb is that a scientist or engineer should be qualified as an expert in his or her major field of scholarship or work and, at the most, two subspecialties. It is incumbent on the attorney, perhaps with the aid of the expert, to find other specialists who can discuss areas in which the expert is not fully conversant. For example, if an attorney approaches a scientist or engineer to testify in a subspecialty of science or engineering in which he or she is not comfortable, it is best for the expert to avoid being talked or flattered into accepting the assignment.

In summary, within the limits of his or her expertise, a testifying scientist or engineer should be able to explain to a jury just what the discipline encompasses. The scientist or engineer, if qualified correctly, should be accepted as being well versed in the aspect of science or engineering relevant to the particular case and, as stated earlier, in perhaps two subspecialties.

The breadth of a scientist's or engineer's knowledge can impress colleagues, but not necessarily juries. In addition to the difficulty of qualifying a scientist or engineer in a number of subspecialties, an attempt to do so may have an adverse effect on the credibility of the scientist or engineer as an expert witness.

If the scientist or engineer does succumb and testifies in too many fields, the opposing counsel will attempt to cast doubt on his or her expertise in cross-examination. It will be easy for the opposing counsel to characterize the expert as a *jack of all trades but a master of none.*

The attorney, as mentioned, should engage other expert witnesses for the presentation of the different fields to be discussed during the trial. If the attorney wants to save money by asking the scientist or engineer expert witness to be all things to all people, it will prove to be a false economy. Alternatively, an attorney retaining a scientist or engineer as an expert in name only, without (for reasons of economy) allowing the expert to read all of the necessary documents, will also prove to be a false economy.

3.2.2 Experience

Experience as a general concept comprises knowledge of, skill in, or observation of something or an event gained through involvement in or exposure to that thing or event. The concept of experience generally refers to know-how or procedural knowledge, rather than propositional knowledge. Philosophers dub knowledge based on experience *empirical knowledge* or *a posteriori knowledge*. A person with considerable experience in a certain field can gain a reputation as an expert.

Experience consists not only of the number of years that a scientist or engineer has been working in his or her chosen discipline (which can be a valuable asset) but also of whether the would-be expert is capable of teaching the subject matter of the case. The judge and jury, being nontechnical and unlikely to be skilled in the subject matter of the case, need to be educated. However, this does not mean that academics make good expert witnesses. Many academics are very poor teachers; in the witness box, if given the opportunity, they tend to ramble, giving opposing counsel the opportunity for a "shark attack" with the resulting destruction of the expert's testimony.

Teaching experience adds a great deal if the expert has taught some phase of science or engineering. The judge and jury will also be interested to know the nature of the teaching and whether or not the expert has taught courses related to the subject matter of the case to undergraduates or graduates, or both.

Practical experience adds further to the credibility of the expert witness. This includes the experience related to the case, the number of years of such experience, and the precise relationship between the case and the work performed by the expert. In addition, scientific or engineering research in the same or similar fields as the subject matter of the case is a benefit to the credibility of the potential expert.

3.2.3 Other Tangible Qualifications

3.2.3.1 Membership in Professional Societies

Another aspect of the expert's credentials that will surely arise is *membership in one or more professional societies*. However, membership in professional societies does not mean that the would-be expert is qualified for the case. It may be pointed out by opposing counsel that any scientist or engineer with minimal qualifications can join one or more professional societies as long as he or she pays the annual dues. Membership in professional societies usually means that the member has reached a certain level of scholarship or that the member pays his or her fees regularly—even on time! In fact, additional qualifications that arise from membership in the society must be vetted carefully because it may be a matter of paying an additional fee and not be related to any additional qualifications.

At this point, the scientific or engineering expert will need to acknowledge membership in and explain the differences between the American Chemical Society (ACS) and the American Institute for Chemical Engineers (AIChE), for example, and how the societies assist the expert in maintaining currency in his or her profession. Membership in the foreign counterpart societies must be explained to show that the foreign societies are not lower on the ladder of recognition than their American counterparts.

3.2.3.2 Publications

It will not be surprising if, at some time during the case—particularly during the trial and usually during cross-examination—the matter of *publications in scientific or engineering journals* will raise its head. The court and the jury will have an interest in the subject matter and the number of publications. Publications that are specifically related to the subject of the court case will enhance the expert's credibility. At some point, the expert will need to explain to a jury the difference between refereed journals and nonrefereed journals.

The issue of publications is often cited as a make-or-break point for any scientist or engineer who has thoughts of being an expert witness. However, it can be a disadvantage. Academics are expected to publish the results of their research. The same is not the case for scientists and engineers working in industry. In the former case, the attorney can evaluate the would-be expert's credibility not only by looking at the number of the publications but also by checking several specific publications for eloquence of the written word and whether or not such a writing style would be effective in a written report to the court. At the same time, looking for publications specifically related to the subject of this court case is a must.

For the industrial scientist or engineer, the attorney may find it difficult to assess scientific and engineering qualifications because publication of data by the industrial scientist or engineer is more difficult. Companies tend to be less enthusiastic about publications in peer-reviewed scientific and engineering journals and opt to publish results as patents, which are written in a specific language by a patent attorney. Although the industrial scientist or engineer will have no doubt about his or her qualifications as they relate to the case, the assessment of this person's abilities by the attorney can be a difficult task.

3.2.4 Nontangible Qualifications

3.2.4.1 Credibility

Credibility is a nontangible qualification that results from qualifications and experience as well as the manner in which the expert presents himself of herself to the judge and jury. Credibility comprises the objective and subjective components of the believability of a scientist or engineer. Traditionally, credibility is composed of two primary dimensions: *trustworthiness* and *expertise*, the order of which can be interchanged and which have both objective and subjective components.

3.2.4.2 Trustworthiness

Trustworthiness is a receiver judgment based on subjective factors. Similarly, *expertise* can be subjectively perceived but includes relatively objective characteristics of the source or message as well (e.g., source credentials or information quality). According to some professional societies, professional *integrity* is the cornerstone of credibility.

Whether a case goes to trial or not, the expert's duties often include reviewing pertinent documents and site conditions as well as evaluating their findings and the developing opinions and conclusions with respect to the case evidence. Thus, the attorney needs to rely upon the expert (trustworthiness) to review all of the pertinent documents. Scientists or engineers cannot present half truths but must give the complete truth, revealing the negatives as well as the positives on each case.

An expert's opinions must be reliable insofar as any opinions and conclusions must be based on generally acceptable standards in the appropriate scientific or engineering milieu. Proof may come from offering up published books on the subject, case-related articles in trade journals, or the testimony of peers.

However, more than simply adhering to standards, experts must be able to fill the gap between the facts that they are examining and their resulting conclusions. It is not sufficient simply to make the connection, say, between his or her opinions and conclusions for the (say) plaintiff. An effective expert witness must be able to show a well-reasoned basis for reaching opinions and conclusions. If the expert uses a theory to this end but is unable to prove that the theory is accepted in the field or unable to intertwine the facts with his or her opinions and conclusions through common-sense reasoning, the judge may (and also may instruct the jurors to) ignore the expert's testimony.

3.2.4.3 Personality

Personality is a nontangible qualification insofar as it is a personal character trait rather than a result of professional training and education in scientific and engineering disciplines. This is an aspect of the would-be expert's qualifications that can only be determined by the attorney by assessing the manner in which the would-be expert will behave in the witness chair under the scrutiny of the opposing counsel. This can be answered, in part, during a presentation made by the would-be expert to the attorney and his or her partners. Poise during the presentation and the ability to field questions can be indicators of the presenter's personality.

The attorney must determine areas of science or personal problems that are uncomfortable for the would-be expert. In fact, the attorney should bring this matter up at the first meeting or soon thereafter and all topics should be aired.

Scientists and engineers can be somewhat reticent in the presence of nontechnical persons (especially attorneys); nevertheless, the attorney should also determine if the would-be expert has any problems with the case about to be tried. As part of this interview, the attorney should also determine if the would-be expert has a "skeleton in the closet" or if any part of the résumé can be questioned. In fact, the would-be expert should inform the attorney as soon as possible during the first contact about any real or potential skeletons in the closet.

Perhaps of more importance to the attorney is whether or not the would-be expert will accept suggestions and even if the would-be expert is willing to sign a report (as his or her own work) that is, in reality, authored by the attorney. The former is grounds for maintaining a relationship; the latter is grounds for terminating any potential relationship because this will come out in court and could cost the attorney the case.

The bottom line is the ability of the expert witness not to cave under cross-examination; through the answers to these questions, the would-be expert might have some ideas of his or her abilities, but the attorney is the best judge of this issue.

In short, a testifying expert should be a person who is able to give the jury a crash course on the subject in an engaging way that holds people's interest. The expert must not appear to be upset or annoyed at the questions (no matter how personal some questions might be) posed by opposing counsel.

3.2.4.4 Appearance of the Expert

Although this issue is addressed in Chapter 7, it is an important intangible factor that is worth mentioning at this point. To recap, the goal of the scientist or engineer as an expert witness is to present the truth to the court by assessing and evaluating the conditions on a case related to his or her individual field of specialty and scholarship.

However, a case can be won or lost depending upon how well the expert witness delivers his or her information in court. It is extremely important that an expert speak with authority to communicate opinions and conclusions to the judge and the jury effectively. From the moment the expert witness takes the stand, the jury members begin to draw their own opinions; therefore, the demeanor and appearance of the expert witness are important in establishing credibility.

The style of dress should be conservative and neat, and the expert should be well groomed. Answers given to questions presented by the attorney under direct examination and by the opposing counsel under cross-examination should be brief and succinct. The answers should convey that the witness is confident, forthright, and professional without the expert's projecting any level of egotistical behavior to the judge and jurors. Most of all, when communicating the answers to questions, the expert should not volunteer information outside his or her area of expertise; such responses will certainly weaken the credibility of the witness.

3.2.5 The Limitations of the Expert

Throughout all of the interchanges between the attorney and the scientist or engineer, the would-be expert must recognize the issues that are at hand and whether or not he or she is equipped for the task. A famous one-liner from one of Clint Eastwood's movies is "A man's got to know his limitations." This is very true for the expert witness, who must also recognize his or her limitations insofar as to know one's *limitations* is to know the reach and limits of one's abilities.

If the scientist or engineer does not know his or her limitations, then being an expert witness can and will be a truly harrowing experience that is not for the fainthearted. If the lack of capabilities of the potential expert witness has gotten beyond the attorney, the opposing counsel, having sharpened his or her teeth during the morning ablutions, will certainly have a field day. As a result, the would-be expert witness will be so ineffective and shown to be lacking ability and credibility to such an extent that a case that has a high probability of a *win* will be turned into a *loss* because of the failure of the would-be expert to recognize his or her own limitations as an expert witness. The issue then becomes a matter of the would-be expert's recognizing his or her qualifications vis-à-vis the case at hand and whether or not his or her credibility will stand the test of time and the verbal onslaught of the opposing counsel.

On the other hand, the attorney who interviews the would-be expert cannot be expected to know all of the quirks and quarks of the expert's background. One type of interview is to invite the expert to the attorney's office to make a presentation to the attorney and any partners on issues related to the background of the case. This will, at least, give the attorney a feeling of the competency of the would-be expert.

However, it all comes down to whether or not the expert recognizes his or her own experience and credibility (i.e., whether the would-be expert recognizes his or her own limitations). In general, the scientist or engineer must have sufficient knowledge of the technical aspects of a case to apply his or her knowledge and form an opinion.

There are real and pseudo-knowledgeable experts, and there are well-intentioned scientists who are not qualified in a particular field. Also available are frauds and charlatans who, for a fee or because they firmly believe that they alone are always right, are willing to testify and to give opinions.

Therefore, the qualifications of the true scientist and engineer become critical to the attorney and, in turn, to the jury, who needs to understand and believe the expert's testimony.

As a start to the process of recognizing one's limitations and perhaps assisting the attorney to determine if the scientist or engineer can truly be an expert witness, a set of criteria must first be investigated by the scientist or engineer and by the attorney who may retain this person. Thus, in order to assist a judge or jury in deciding a matter in dispute, an expert must have sufficient knowledge, training, or experience in the field in which he or she will testify. If the issue involves analyzing the actions of a licensed professional, for example, the testifying expert should at least be licensed in the same technical area and have the same experience as, if not more experience than, the experts retained by the opposing counsel.

Whether an expert has previously testified in front of a jury or judge is not as critical an element as his or her professional experience and credibility are. It is not sufficient for an expert merely to say he or she has the proper training and experience. Expertise must be proven with hard evidence. This is done by showing educational degrees, licenses, work history, membership in professional organizations, accomplishments, published articles in trade journals, and awards or recognition by others in and out of the field. If the expert cannot back up his declared expertise, he or she will most likely be barred from testifying.

3.3 THE CLIENT

Simply, the client is the person or organization that employs and retains a scientist or engineer as an expert witness to produce expert testimony in a case. The client is also the person or the organization that retains the attorney or counselor to advise about some legal matters or to manage or defend a suit or action in which the client is a party. Thus, the client is the party for which professional services are rendered by the scientist or engineer—either the attorney or the disputant (plaintiff or defendant).

The duties of the client toward his or her counsel include: (1) giving him or her written authority, (2) disclosing his or her case with perfect candor, (3) offering advances of money to his or her attorney spontaneously, and (4) paying his or her attorney fees promptly at the conclusion of the suit or matter.

On the other hand, the client's rights include: (1) being diligently served in the management of his or her business, (2) being informed of the progress of the case, and (3) expecting that his or her counsel will not disclose what has been professionally confided.

It is expected that the scientist or engineer, who is often contacted by the attorney rather than being contacted directly by the client, should offer the same duties and rights to the client and the attorney.

There are a number of different types of clients who may require the services of an expert witness:

Attorneys may approach a scientist or engineer for service, but they may not be the client. Instead, they may be the indirect client, working directly for a disputant. In some instances the attorney will be retained by an insurance company, and the insurer normally pays all costs of defense, including the fees of the scientist or engineer.

An *insurance company* may be the direct client, which sometimes occurs when it has been advised of an incident and assumes a claim will be filed. In those circumstances the company wants to obtain technical consultation to prepare for the claim which almost assuredly will follow. In other cases, the insurer may have already paid on an insured's claim and looks to pursue the causal party to recover payment by obtaining rights of its insured through subrogation. In either case, the expert usually winds up working with an attorney.

Public entities will sometimes wish to engage an expert to provide high-level expertise relative to various matters ranging from road construction through an environmentally sensitive area to companies taking a tax credit that is being reviewed and questioned by the relevant government department.

Individuals often retain an expert to testify at public hearings of different types, often to support or oppose the position of a public entity. In some cases, an individual will seek to pursue a matter, often against the advice of counsel.

Courts in some jurisdictions have the authority to retain an expert. Payment in such cases may be fixed by the court. Payment is made from the jurisdiction's treasury or is apportioned among the litigants.

3.4 THE INITIAL ATTORNEY–EXPERT CONTACT

Often the scientist or engineer is called by an attorney who will inquire as to his or her willingness and availability to serve as an expert. In many cases the potential expert will be told that the attorney's client has been wronged and this requires an effective expert witness to prove this to be the case. If there has already been a call relative to that same case and the scientist or engineer has been retained by another attorney for the case, the would-be expert should thank the attorney for the call and immediately decline to serve. The attorney who has engaged the would-be expert should be told of the call. In addition, a number of issues can be resolved during the initial telephone call to determine whether or not it is in the best interests of either party to spend any more time with one another.

3.4.1 The Initial Contact

In the beginning, the scientist or engineer should attempt to learn something about the dispute, such as the identity of the plaintiff, the allegations, the defendant and his or her position, and whether a countersuit exists. Then, given the nature of the dispute, it is a question of the type of expertise the attorney is seeking. If the scientist or engineer does not have that expertise or has never been retained as an expert witness, the attorney should be told immediately. Moreover, the would-be expert needs to determine if the attorney is expecting him or her to take a preconceived position. If this is disagreeable to the scientist or engineer, it is best to decline politely and sever all communication with the attorney. Alternatively, if the attorney determines that the would-be expert has been contacted by opposing counsel and is taking one side of the case or the other for a higher fee, the attorney should end the conversation and consider whether or not to report this to the court at the appropriate time.

3.4.2 Conflict of Interest

To determine if there may be any potential conflict of interest, the scientist or engineer should ask about the composition of the opposing team. If the scientist or engineer has been retained by the opposing counsel on a previous case, the attorney now calling should be so informed. If he or she has worked in the past for the opposing counsel's client, that, too, should be made known to the attorney. The same applies to others involved with the opposing attorney's client.

A scientist or engineer about to be retained as an expert on a case must not only appear objective, but must be objective. If any type of prior relationship has existed with any of the principals involved—even peripherally—a conflict may be alleged and the expert's presentation discredited and credibility destroyed.

3.4.3 Attorney–Expert Relationship

Very early in the relationship between the attorney and the expert, the discussion should include areas of science or personal problems that are uncomfortable for the expert. In fact, the attorney should bring this matter up at the first meeting or soon thereafter. In addition to topics discussed under the preceding section, all other topics should be aired. These can include whether the expert (1) has any problems with the fee, (2) will not hesitate to tell the jury what reimbursements are expected, (3) has any problems with the case about to be tried, (4) has a fear that he or she will not make a good witness on the stand, or (5) has a "skeleton in the closet" insofar as his or her résumé can be questioned and he or she can be thrown into disrepute and credibility disappears. As a final element of discussion, the expert and the attorney should set a date for a meeting but allow enough time to perform research into their respective backgrounds.

Principal issues for the would-be expert relate to the attorney's standing with his or her colleagues and his or her thoroughness of trial preparation. Other issues include the attorney's participation in the expert's work and whether or not he or she attempts to learn more about the technical issues to develop and rehearse testimony. The same types of questions could be asked of other experts the attorney has worked with in the past, as well as any others he has engaged for the present case. If the attorney has for some reason dismissed an expert, by all means the would-be expert should contact the individual to learn more about what happened. It may be more than a matter of the expert's being unable to support the attorney's point of view. Other issues that merit inquiry include the degree of research freedom the attorney permits and whether he or she discloses all information to experts.

The would-be expert may also feel the need to determine the maturity of the case. If the case is just getting under way, there may be no issues. However, if the case is about to go to trial, the scientist or engineer will need to know the disposition of any prior experts. If the lawyer indicates there was one but is reluctant to give his or her name, there is cause for concern. On the other hand, if there have been no prior experts and the lawyer has delayed this long, it may indicate that he or she is careless or operating on a shoestring; generally speaking, this is the kind of lawyer who often loses cases that should be won.

Principal issues for the attorney relate to the reliability of the would-be expert in terms of his or her standing with other attorneys, especially his or her thoroughness of trial preparation. Other issues include the expert's participation in the work and whether or not he or she attempts to learn more about the technical issues of the case as they affect the would-be expert's portion of the case. The attorney may wish to consult trial transcripts of other cases on which the scientist or engineer has worked to determine eloquence of writing, how evidence is presented, stability under cross-examination, and the potential for team work. If another attorney has previously dismissed the would-be expert, the attorney needs to determine for what reason and, if necessary, sever communication. Other issues that merit inquiry by the attorney include the would-be expert's billing practices and whether or not he or she disclosed all information to previous attorneys.

3.5 THE EXPERT AND THE DISPUTE

The function of the scientist or engineer as an expert witness is to present evidence related to the technical aspects of the case. *Evidence* is a collection of facts and statements admitted on the record of the trial and made known to the court or jury that tends to prove or disprove an issue or dispute. Therefore, evidence can be broadly defined as any item introduced during the course of a trial for the purpose of establishing some factual proposition that is in controversy in the lawsuit. Documents or physical items and testimony may be offered and either admitted or not admitted. If not admitted, the material or testimony is not evidence on the record and may not be considered in deciding the issue before the court. If admitted, sometimes it carries great weight and, at other times, it carries little weight.

3.5.1 Types of Evidence

There are two basic types or categories of evidence: (1) direct and (2) circumstantial. *Direct evidence* (sometimes called *eye-witness evidence*) is something that proves a factual proposition without resorting to logical inference—such as what a witness saw, heard, smelled, tasted, or felt. *Circumstantial evidence* (or *indirect evidence*) consists of facts from which certain logical inferences must be drawn in order to establish or disprove another fact relevant to the lawsuit.

In addition to the types of evidence, there are two forms of evidence: (1) physical evidence and (2) testimonial evidence. *Physical* (*real* or *tangible*) *evidence* is the evidence that is marked and entered. It is something that can be picked up and examined, such as (in the current context) charts, summaries, laboratory reports, scientific documents, and an expert witness report. It is introduced through persons who are called upon to testify, such as an expert witness report. *Testimonial evidence* may be the statement of an expert witness under oath during the course of the trial. The report submitted by an expert witness is *tangible evidence,* but the explanation of the report by the expert from the witness stand is *testimonial evidence.*

For evidence to be received by a court during the course of a trial, the evidence must be of a probative material factual issue and be otherwise competent under the rules of evidence:

Probative evidence tends to either prove or disprove some factual controversy in the case.

Material evidence is examined at the pleadings (or the indictment) to see what is really at issue, but the evidence must be relative.

Relevant evidence is defined by Rule 401 of the Federal Rules of Evidence and is evidence that has the tendency to make the existence of any fact more probable or less probable than it would be without the evidence. The *relevant evidence rule* holds that only evidence that affects issues under dispute, as outlined in the pleadings, may be used. It is particularly germane when considering circumstantial evidence, which constitutes indirect proof or disproof of a fact in question. Even relevant evidence can be excluded through the *hearsay rule,* whereby a witness cites what someone else told him or her.

Incompetent evidence is evidence that may be relevant and probative but may not be admitted because it is not offered in the proper form. The most common example of incompetent evidence is hearsay testimony in which a witness is allowed to testify concerning what he or she was told by another party. The person who told the witness is relevant to the issue and should be on the witness stand under oath. Other common examples of incompetent evidence are (1) secondary evidence of written material, (2) unfairly prejudicial evidence, and (3) privileged communications.

The *parole evidence rule* makes inadmissible any evidence of understandings different from those formally entered into (as in the case of an existing contract) when the formal agreement was established after such understandings were alleged to exist.

3.5.2 Hearsay Evidence

Hearsay is an out-of-court assertion offered to prove the matter asserted. An assertion or assertive conduct may be oral or something that is said, but it may also be waving the hands, raising the eyebrows, or some other bodily movement. Hearsay is particularly of interest to the scientist or engineer who tends to discuss the result of his or her work with peers. Statements are made that stick in the mind and may be verbally regurgitated at some appropriate or inappropriate point. However, in the court system, hearsay has a broad convergence because it includes oral statements, documentary evidence, assertive conduct, and, in certain circumstances, even the prior statements of the witness testifying. It is at the discretion of the court whether hearsay is allowed or disallowed.

The expert should assume, for all intents and purposes, that hearsay is not admissible. There are exceptions, however, provided by the Federal Rules of Evidence (Table 3.1) and by other rules prescribed by the Supreme Court and by acts of Congress. If hearsay is admissible, the adverse party would be deprived of the opportunity for cross-examination to test the expert's opinions and conclusions. The objective of the cross-examination is to ensure the trustworthiness of the evidence.

The exceptions to the hearsay rule, within the present context, include (1) prior statements of a witness and (2) admission by a party-opponent. The *prior statement of a witness* is a statement that (1) is inconsistent with the declarant's present testimony and was given under oath and (2) is consistent with the declarant's present testimony and is offered to rebut a charge (of recent fabrication) against the declarant. *Admission by a party-opponent* is a statement offered against a party that (1) is the party's own statement, (2) is a statement that the party has manifested as an adoption or belief in its truth, (3) is a statement by a person authorized by the party to make a statement concerning the subject, or (4) is a statement made by the party's authorized agent serving in the capacity of the party's authorization.

TABLE 3.1
Federal Rules of Evidence as They Apply to the Scientist or Engineer as an Expert Witness

Federal Rules of Evidence

Rule 601. General Rule of Competency

Every person is competent to be a witness except as otherwise provided in these rules. However, in civil actions and proceedings, with respect to an element of a claim or defense as to which State law supplies the rule of decision, the competency of a witness shall be determined in accordance with State law.

Rule 602. Lack of Personal Knowledge

A witness may not testify to a matter unless evidence is introduced sufficient to support a finding that the witness has personal knowledge of the matter. Evidence to prove personal knowledge may, but need not, consist of the witness' own testimony. This rule is subject to the provisions of rule 703, relating to opinion testimony by expert witnesses.

Rule 603. Oath or Affirmation

Before testifying, every witness shall be required to declare that the witness will testify truthfully, by oath or affirmation administered in a form calculated to awaken the witness' conscience and impress the witness' mind with the duty to do so.

Rule 608. Evidence of Character and Conduct of Witness

(a) Opinion and reputation evidence of character

The credibility of a witness may be attacked or supported by evidence in the form of opinion or reputation, but subject to these limitations:

1. the evidence may refer only to character for truthfulness or untruthfulness, and
2. evidence of truthful character is admissible only after the character of the witness for truthfulness has been attacked by opinion or reputation evidence or otherwise.

(b) Specific instances of conduct

Specific instances of the conduct of a witness, for the purpose of attacking or supporting the witness' credibility, other than conviction of crime as provided in rule 609, may not be proved by extrinsic evidence. They may, however, in the discretion of the court, if probative of truthfulness or untruthfulness, be inquired into on cross-examination of the witness

1. concerning the witness' character for truthfulness or untruthfulness, or
2. concerning the character for truthfulness or untruthfulness of another witness as to which character the witness being cross-examined has testified.

The giving of testimony, whether by an accused or by any other witness, does not operate as a waiver of the accused's or the witness' privilege against self-incrimination when examined with respect to matters which relate only to credibility.

Rule 611. Mode and Order of Interrogation and Presentation

(a) Control by court

The court shall exercise reasonable control over the mode and order of interrogating witnesses and presenting evidence so as to

1. make the interrogation and presentation effective for the ascertainment of the truth,
2. avoid needless consumption of time, and

TABLE 3.1 (CONTINUED)
Federal Rules of Evidence as They Apply to the Scientist or Engineer as an Expert Witness

3. protect witnesses from harassment or undue embarrassment.

(b) Scope of cross-examination

Cross-examination should be limited to the subject matter of the direct examination and matters affecting the credibility of the witness. The court may, in the exercise of discretion, permit inquiry into additional matters as if on direct examination.

(c) Leading questions

Leading questions should not be used on the direct examination of a witness except as may be necessary to develop the witness' testimony. Ordinarily, leading questions should be permitted on cross-examination. When a party calls a hostile witness, an adverse party, or a witness identified with an adverse party, interrogation may be by leading questions.

Rule 612. Writing Used to Refresh Memory

Except as otherwise provided in criminal proceedings by section 3500 of title 18, United States Code, if a witness uses a writing to refresh memory for the purpose of testifying, either

1. while testifying, or
2. before testifying, if the court in its discretion determines it is necessary in the interests of justice,

an adverse party is entitled to have the writing produced at the hearing, to inspect it, to cross-examine the witness thereon, and to introduce in evidence those portions which relate to the testimony of the witness. If it is claimed that the writing contains matters not related to the subject matter of the testimony, the court shall examine the writing in camera, excise any portions not so related, and order delivery of the remainder to the party entitled thereto. Any portion withheld over objections shall be preserved and made available to the appellate court in the event of an appeal. If a writing is not produced or delivered pursuant to order under this rule, the court shall make any order justice requires, except that in criminal cases when the prosecution elects not to comply, the order shall be one striking the testimony or, if the court in its discretion determines that the interests of justice so require, declaring a mistrial.

Rule 613. Prior Statements of Witnesses

(a) Examining witness concerning prior statement

In examining a witness concerning a prior statement made by the witness, whether written or not, the statement need not be shown nor its contents disclosed to the witness at that time, but on request the same shall be shown or disclosed to opposing counsel.

(b) Extrinsic evidence of prior inconsistent statement of witness

Extrinsic evidence of a prior inconsistent statement by a witness is not admissible unless the witness is afforded an opportunity to explain or deny the same and the opposite party is afforded an opportunity to interrogate the witness thereon, or the interests of justice otherwise require. This provision does not apply to admissions of a party-opponent as defined in rule 801(d)(2).

Rule 614. Calling and Interrogation of Witnesses by Court

(a) Calling by court

The court may, on its own motion or at the suggestion of a party, call witnesses, and all parties are entitled to cross-examine witnesses thus called.

TABLE 3.1 (CONTINUED)
Federal Rules of Evidence as They Apply to the Scientist or Engineer as an Expert Witness

(b) Interrogation by court

The court may interrogate witnesses, whether called by itself or by a party.

(c) Objections

Objections to the calling of witnesses by the court or to interrogation by it may be made at the time or at the next available opportunity when the jury is not present.

Rule 615. Exclusion of Witnesses

At the request of a party, the court shall order witnesses excluded so that they cannot hear the testimony of other witnesses, and it may make the order of its own motion. This rule does not authorize exclusion of

1. a party who is a natural person, or
2. an officer or employee of a party which is not a natural person designated as its representative by its attorney, or
3. a person whose presence is shown by a party to be essential to the presentation of the party's cause.

Rule 701. Opinion Testimony by Lay Witnesses

If the witness is not testifying as an expert, the witness' testimony in the form of opinions or inferences is limited to those opinions or inferences which are (a) rationally based on the perception of the witness and (b) helpful to a clear understanding of the witness' testimony or the determination of a fact in issue.

Rule 702. Testimony by Experts

If scientific, technical, or other specialized knowledge will assist the trier of fact to understand the evidence or to determine a fact in issue, a witness qualified as an expert by knowledge, skill, experience, training, or education may testify thereto in the form of an opinion or otherwise.

Rule 703. Bases of Opinion Testimony by Experts

The facts or data in the particular case upon which an expert bases an opinion or inference may be those perceived by or made known to the expert at or before the hearing. If of a type reasonably relied upon by experts in the particular field in forming opinions or inferences upon the subject, the facts or data need not be admissible in evidence.

Rule 704. Opinion on Ultimate Issue

(a) Except as provided in subdivision (b), testimony in the form of an opinion or inference otherwise admissible is not objectionable because it embraces an ultimate issue to be decided by the trier of fact.

(b) No expert witness testifying with respect to the mental state or condition of a defendant in a criminal case may state an opinion or inference as to whether the defendant did or did not have the mental state or condition constituting an element of the crime charged or of a defense thereto. Such ultimate issues are matters for the trier of fact alone.

Rule 705. Disclosure of Facts or Data Underlying Expert Opinion

The expert may testify in terms of opinion or inference and give reasons without first testifying to the underlying facts or data, unless the court requires otherwise. The expert may, in any event, be required to disclose the underlying facts or data on cross-examination.

TABLE 3.1 (CONTINUED)
Federal Rules of Evidence as They Apply to the Scientist or Engineer as an Expert Witness

Rule 706. Court-Appointed Experts

(a) Appointment

The court may, on its own motion or on the motion of any party, enter an order to show cause why expert witnesses should not be appointed, and may request the parties to submit nominations. The court may appoint any expert witnesses agreed upon by the parties, and may appoint expert witnesses of its own selection. An expert witness shall not be appointed by the court unless the witness consents to act. A witness so appointed shall be informed of the witness' duties by the court in writing, a copy of which shall be filed with the clerk, or at a conference in which the parties shall have opportunity to participate. A witness so appointed shall advise the parties of the witness' findings, if any; the witness' deposition may be taken by any party; and the witness may be called to testify by the court or any party. The witness shall be subject to cross-examination by each party, including a party calling the witness.

(b) Compensation

Expert witnesses so appointed are entitled to reasonable compensation in whatever sum the court may allow. The compensation thus fixed is payable from funds which may be provided by law in criminal cases and civil actions and proceedings involving just compensation under the Fifth Amendment. In other civil actions and proceedings the compensation shall be paid by the parties in such proportion and at such time as the court directs, and thereafter charged in like manner as other costs.

(c) Disclosure of appointment

In the exercise of its discretion, the court may authorize disclosure to the jury of the fact that the court appointed the expert witness.

(d) Parties' experts of own selection

Nothing in this rule limits the parties in calling expert witnesses of their own selection.

3.5.3 The Best-Evidence Rule

The *best-evidence rule* falls under the Federal Rules of Evidence, Section 1002, and is related to proving the content of a writing, recording, or photograph; the original is required, unless some exception (Rules of Evidence or Act of Congress) applies. However, under the Federal Rules of Evidence, Section 1003, a duplicate is admissible to the same extent as an original unless (1) a genuine question is raised as to the authenticity of the original or (2) in the circumstances, it would be unfair to admit the duplicate in lieu of the original. For the purposes of this rule, a duplicate may be a carbon copy of a charge slip or photocopies of checks of original microfilm in the absence of a suggestion that the photocopies are incorrect.

Then, under the Federal Rules of Evidence, Section 1004, the original is not required, and other evidence of the contents of the writing, recording, or photograph is admissible if: (1) all of the originals have been lost or have been destroyed, unless the proponent lost or destroyed them in bad faith; (2) no original can be obtained by any available judicial process or procedure; (3) at a time when an original was under the control of the party against whom it is offered, he or she was put on notice, by the pleadings or otherwise, that the contents would be a subject of proof at the hearing, and he or she does not produce the original at the hearing; or (4) the writing, recording, or photograph is not closely related to a controlling issue.

Finally, under the Federal Rules of Evidence, Section 1004, the contents of voluminous writings, recordings, or photographs that cannot be conveniently examined in court may be presented in the form of a chart, summary, or calculation. However, the originals or duplicates must be available for examination or copying, or both, by other parties at a reasonable time and place. The court may order that they are to be produced in court. It is not necessary to prove that it is impossible to examine the documents because they are too voluminous, but only that it is inconvenient.

Thus, evidence (either testimonial or tangible) must be competent and probative of the factual issues in the case. The evidence must be sponsored by an individual with personal knowledge about it. If testimonial evidence occurred outside the courtroom, it is hearsay evidence and cannot be introduced in court unless an exception to the hearsay rule is applicable. Finally, all tangible evidence must be authenticated. Therefore, it might be suggested that the best form of tangible evidence is for the plaintiff to use the defendant's documented statements against him or her.

3.5.4 Federal Rules of Evidence

The expert is not usually (or is only very rarely) an attorney; however, understanding of the Federal Rules of Evidence, which prescribe the substance to which witnesses may testify, can be helpful. These rules describe the circumstances under which a witness may give an opinion. If the scientist or engineer has some knowledge of the Federal Rules of Evidence, he or she will be better able to understand why the argument occurs and have greater confidence that when the Court rules, he or she will be allowed to testify and the materials that the expert may use in preparing his or her reports will be acceptable. It is important first to understand what testimony may be given by a lay witness so that the difference may be seen in the type of testimony an expert witness may give and the importance of qualifying an expert witness to testify in certain situations.

> *Rule 701. Opinion Testimony by Lay Witnesses.* If a witness is not testifying as an expert, the witness's testimony in the form of opinions or inferences is limited to those opinions or inferences that are (1) rationally based on the perception of the witness, and (2) helpful to a clear understanding of the witness's testimony or the determination of a fact in issue.

The rule implies that a judge would exclude opinion evidence if it is not helpful in clarifying the facts or if it is confusing. The rule assumes that the natural characteristics of the adversary system will generally lead to an acceptable result because the detailed account carries more conviction than the broad assertion, and a lawyer can be expected to display his or her witness to the best advantage. If he or she fails to do so, cross-examination and argument will point up weaknesses. If, despite these considerations, attempts are made to introduce meaningless assertions that amount to little more than choosing sides, exclusion for lack of helpfulness is called for by the rule.

In order to conclude that lay witness opinion testimony is admissible, the court must find that the witness's testimony is based upon personal observation and recollection of concrete facts and that those facts, without the opinion, cannot be described in sufficient detail to convey the substance of the testimony to the jury adequately.

> *Rule 702. Testimony by Experts.* If scientific, technical, or other specialized knowledge will assist the trier of fact in understanding evidence or determining a fact in issue, a witness qualified as an expert by knowledge, skill, experience, training, or education may testify thereto in the form of an opinion or otherwise.

The requirements are that (1) the testimony will assist the trier of fact in understanding the evidence; (2) an expert may be qualified in scientific, technical, or other specialized areas; (3) an expert may be qualified by reason of knowledge, skill, experience, training, or education; or (4) testimony may be opinion, dissertation on scientific principles, with inferences to be drawn by applying specialized knowledge to facts, or hypothetical questions may be pursued, but not necessarily.

Preliminary questions concerning qualifications of an expert are determined by the court. In making the decision, the court is not bound by the Federal Rules of Evidence, except those with respect to privilege [Rule 104(a)]. This means that the judge may consider hearsay and other inadmissible information in making the decision about whether a witness may testify as an expert.

An evaluation of facts is often difficult or impossible without the application of scientific, technical, or other specialized knowledge. The most common source is the knowledge of an expert. Whether the situation is a proper one for the use of expert testimony is to be determined on the basis of assisting the trier of fact. The most certain test for determining when experts may be used is the "common sense" inquiry—deciding whether the untrained layman would be qualified to determine intelligently and to the best possible degree the particular issue without enlightenment from those having a specialized understanding of the subject involved in the dispute. When opinions are excluded, it is because they are not helpful and are therefore superfluous and a waste of time.

The fields of knowledge that may be drawn upon are not limited merely to the scientific and technical but, rather, extend to all specialized knowledge. Within the scope of the rule, experts may be (in addition to physicians, physicists, petroleum engineers, geologists, architects, and economists) bankers, land owners, mechanics, and building contractors.

> *Rule 703. Bases of Opinion Testimony by Experts.* The facts or data in the particular case upon which an expert bases an opinion or inference may be those perceived by or made known to the expert at or before the hearing. If of a type reasonably relied upon by experts in the particular field in forming opinions or inferences upon the subject, the facts or data need not be admissible in evidence.

The sources of facts include: (1) firsthand knowledge with opinions based thereon, which are traditionally allowed; (2) presentation of facts at trial with the expert in attendance or by hypothetical question; and (3) facts made known to the expert outside the courtroom, which includes graphs, diagrams, maps, reports, and other such materials.

If it is clear that enlargement of permissible data may tend to break down the rules of exclusion (of inadmissible evidence) unduly, notice should be taken that the rule requires that the facts or data *be of a type reasonably relied upon by experts in the particular field.*

The facts to be used in determining the reasonableness of an expert's reliance on inadmissible information in forming his or her opinion in cases outside the "mainstream" of this rule are the extent to which (1) the opinion is pervaded by reliance on materials judicially determined to be inadmissible on grounds of relevance or trustworthiness; (2) the opinion is pervaded by reliance upon other materials; (3) the expert's assumptions have been shown to be unsupported, speculative, or demonstrably incorrect; (4) the materials relied on are within immediate sphere of expertise, are of a kind customarily relied upon by experts in the field, and are not used for litigation purposes only; (5) the expert acknowledges the questionable reliability of underlying information; and (6) reliance on certain materials, even if otherwise reasonable, may be unreasonable in the peculiar circumstances of the case.

> *Rule 704. Opinions on Ultimate Issue.* Testimony in the form of an opinion or inference otherwise admissible is not objectionable because it embraces an ultimate issue to be decided by the trier of fact. In fact, no expert witness testifying with respect to the mental state or condition of a defendant in a criminal case may state an opinion or inference as to whether the defendant did or did not have the mental state or condition constituting an element of the crime charged or of a defense thereto. Such ultimate issues are matters for the trier of fact alone.

Prior to adoption of the Federal Rules of Evidence (January 2, 1975), older cases often contained strictures against allowing witnesses to express opinions upon ultimate issues as a particular aspect of the rule against opinions. The rule was restrictive, difficult to apply, and generally served only to deprive the trier of fact of useful information.

The basic approach to lay opinions and expert opinions in these rules is to admit them when helpful to the trier of fact. In order to render this approach fully effective and to allay any doubt on the subject, the so-called *ultimate issue* rule is specifically abolished by the rule.

> *Rule 705. Disclosure of Facts or Data Underlying Expert Opinion.* The expert may testify in terms of opinion or inference and give reasons therefore without prior disclosure of the underlying facts or data, unless the court requires otherwise. The expert may in any event be required to disclose the underlying facts or data on cross-examination.

This rule eliminates the need for long, hypothetical questions, which have been the target of a great deal of criticism as encouraging partisan bias, as affording an opportunity for summing up in the middle of the case, and as complex and time consuming. Although the rule allows counsel to make disclosure of the underlying facts or data as a preliminary to the giving of an expert opinion if he or she chooses, the instances in which he or she is required to do so are reduced. This is true whether the expert bases his or her opinion on data furnished as secondhand or observed by him or her firsthand.

Under this rule, the facts upon which the opinion is based need not be disclosed prior to trial. The best practice is for all facts to be disclosed in the report. The modern view in evidence law recognizes that experts often rely on facts and data supplied by third parties, and these rules codify an approach permitting disclosure of otherwise hearsay evidence for the purpose of illustrating the basis of an expert witness's opinion.

> *Rule 706. Court-Appointed Experts.* The court may on its own motion or on the motion of any party enter an order to show cause as to why the expert witnesses should not be appointed and may request the parties to submit nominations. The court may appoint any expert witnesses agreed upon by the parties and may appoint expert witnesses of its own selection. An expert witness shall not be appointed by the court unless the witness consents to act. A witness so appointed shall be informed of the witness's duties by the court in writing, a copy of which shall be filed with the clerk, or at a conference in which the parties shall have opportunity to participate. A witness so appointed shall advise the parties of the witness's findings, if any; the witness's deposition may be taken by any party; and the witness may be called to testify by the court or any party. The witness shall be subject to cross-examination by each party, including a party calling the witness.

The ever present possibility that the judge may appoint an expert in a given case must inevitably exert a sobering effect on the expert witness of a party and upon the person utilizing his or her services.

> *T.C. Rule 143(f).* Expert Witness Reports (amended effective 7/1/90). Unless otherwise permitted by the Court upon timely request, any party who calls an expert witness shall cause that witness to prepare a written report for submission to the Court and to the opposing party. The report shall set forth the qualifications of the expert witness and shall state the witness's opinion and the facts or data on which

> that opinion is based. The report shall set forth in detail the reasons for the conclusion, and it will be marked as an exhibit, identified by the witness, and received in evidence as the direct testimony of the expert witness, unless the Court determines that he or she is not qualified as an expert. Additional direct testimony with respect to the report may be allowed to clarify or emphasize matters in the report, to cover matters arising after the preparation of the report, or otherwise at the discretion of the Court.
>
> If not furnished earlier, each party who calls any expert witness shall furnish to each other party and shall submit to the Court, not later than 30 days prior to the call of the trial calendar on which the case shall appear, a copy of all expert witness reports prepared.

The court ordinarily will not grant a request to permit an expert witness to testify without a written report where the expert witness's testimony is based on third-party contacts, statistical data, or other detailed technical information.

According to present procedure, the court issues a trial calendar, listing all of the cases being considered at one time. Many of the cases will be settled before the calendar call.

The primary requirement for opinion testimony is that it be helpful to the trier of fact. Lay witness opinion is limited to those rationally based on perception of the witness. An expert (scientist, engineer, or other specialized professional) is qualified by knowledge, skill, experience, training, or education. The expert's opinion is based on facts or data perceived or provided. Furthermore, an expert may base his or her opinion on hearsay. Sources must be the type normally relied upon in the field by persons specializing in that field; the expert's testimony must be helpful to the court, and the expert must be qualified to give an opinion.

In summary, an expert makes a report, discloses facts, and furnishes the report to the other side (through the expert's counsel because direct contact between the expert and opposing counsel is not recommended) when ordered by the court or at least thirty days prior to the trial.

3.5.5 Assembling Evidence

In order to assemble evidence, the expert should investigate every aspect of the case by reading all of the written material that the attorney wishes to have revealed. The expert must evaluate the weaknesses as well as the strengths of the scientific case being prepared by the attorney. It is incumbent on the expert to inform the attorney of the defects (if any) of the case under consideration; this should only be done orally unless otherwise requested. Usually, the attorney knows the strengths of the case. The expert must be confident when this topic is discussed. If possible, the expert should have specific scientific references to share with the attorney on both the strengths and the weaknesses of the case. It is wise to discuss whether the expert will be asked to point out the strengths or weaknesses of a case and, as always, the expert must remember to be an objective evaluator. Again, all written material, whether it is a formal report or written document or notes, and possibly much of the conversation between the attorney and expert can be subject to discovery.

Discovery is the disclosure of evidence by one party to the other party. In civil cases, the courts encourage and require the parties to disclose the nature of their cases and their anticipated evidence. This is intended to encourage settlements and, if the dispute goes to trial, leads to speedier trials. In criminal cases, the plaintiff (usually the government) must disclose virtually the entire case to the defendant to allow the defendant a right of due process, but the defendant, because of the privilege against self-incrimination, is not required to provide information to the plaintiff. Discovery includes depositions, interrogatories, and requests for production of documents. On the other hand, *stipulations*, unlike discovery (sometimes the two are confused by experts), are merely agreements of the parties as to certain facts. Stipulations speed the trial process and avoid the delay of arguing over details that are not reasonably in dispute.

Thus, discovery essentially means that the expert should not write anything to the attorney without a specific request by the attorney to do so. This advice does not usually apply to the consultant or the nontestifying expert.

3.5.6 The Opinion of the Expert Witness

The expert must review the facts of the case in great detail. A literature search is normally in order, and many scientific and engineering databases are available. However, the search strategy must be carefully thought out to yield only the most relevant information.

During this time, many personal meetings or telephone contacts between the expert and the attorney are absolutely necessary. It is during these contacts that the expert should give the attorney the benefit of the expert's evaluation of the case. All this should be done verbally; however, the expert would be wise to keep a written log of time spent (and what was done) to formulate the opinion.

Only after a request should the expert tell the attorney the opinion or conclusion he or she has derived from the information available. A subtle point here is that the expert is not engaged by the attorney to give an opinion but, rather, to *derive* an opinion from the facts of the case. The expert must be assured and be careful that the attorney is not trying to buy an opinion.

Sometimes an expert writes a preliminary draft of the opinion. If so, the attorney can help the expert state his or her opinion in legal language or any other terms, but only if the attorney thinks it is necessary. Under no circumstances should the expert permit the attorney to put words in his or her mouth, the nature of which may slant the opinion.

A copy of the opinion may be sent to the opposing counsel as a result of the discovery process. If the expert is not expected to go to trial, his or her opinion can be limited and may not be revealed for discovery. The written report stating the opinion in clear words is one of the most important documents because it can help the attorney decide to settle or drop the case or take it to court.

If at any time during the case the expert is asked if he or she has an opinion, the response should be a simple one-word answer: *yes* or *no*. It is not necessary for the expert to elaborate on his or her answer unless he or she is under oath in a deposition or in court. When asked about his or her opinion, the expert should make a simple statement, of perhaps one sentence, for entry into the deposition transcript or to give the court the benefit of the conclusion reached. Again, it is not necessary to elaborate at this point. The attorney may then ask whether the expert believes the factual premises are true and if the expert's opinion is based on them. Again, a simple *yes* or *no* should suffice. The attorney should ask the reason or basis (or some such word) for the opinion. Only then should the expert go on as long as it takes to explain the details of the facts and ideas behind the opinion. Too many experts give too many details at first without considering the consequences of giving away the story at an early stage.

3.5.7 Scope of Services

It will be difficult to pinpoint the scope of services at the beginning. Some will become known only after the work progresses. Accordingly, it is suggested that initial services be identified by the attorney and the expert and that the option remain for other services to be included at a later date. In this respect, the contract should include a list of the services the expert will provide, indicating also any special rates that may apply. For example, most research services will be applied at a standard rate. However, the fees associated with giving a deposition or testimony in court may be higher, to account for the stress that may be involved. Conversely, on-call services may be some fraction of the standard rate. Some experts are uncomfortable charging different rates for different tasks and prefer a standard hourly rate with the caveat: *I work, you pay; I do not work, you do not pay.*

Many an opposing counsel has had a field day when cross-examining an expert who charges different rates for different tasks. When under such cross-examination, many experts have sat and squirmed in the proverbial hot seat. Jurors may look askance at practices that use different rates and wonder about the credibility and impartiality of the expert.

Some experts charge a fee based on the amount of money at stake in the case; usually, the fee is calculated as a percentage of the settlement. Such experts appear to be advocates more than impartial professional experts. Knowing that the fee (and payment) depends upon the outcome of the case, judges and jurors are not too certain of the credibility of such experts.

3.6 TIME TO WALK AWAY

Once the issues outlined in the previous section have been addressed, it is then possible for the scientist or engineer to make a self-assessment and determine if he or she is qualified to be an expert witness. The attorney can make a similar assessment and reach a decision knowing that, at some point of the proceedings, the opposing counsel will perform a background check on the expert witness and form his or her own conclusion that will undoubtedly be aired in court.

No matter how strongly the would-be expert feels about his or her qualifications and the chance to be retained as an expert witness, he or she must be of a mind to examine himself or herself carefully in light of the issues in the case. If there is any doubt, it may be time to walk away from this process when he or she recognizes there is no possibility of making a meaningful contribution to the case. If not, it may be proven in court that he or she is not capable of gaining the respect of the judge and the jury. The attorney must also decide whether or not retaining this particular scientist or engineer is appropriate for the case.

If there is a match between the qualifications of the would-be expert and the subject matter of the case, it is time to proceed on a positive note. If not, it is time for both parties to walk away!

3.7 CONFIDENTIALITY AND NONDISCLOSURE

The would-be expert should always indicate that he or she is ready to sign a confidentiality agreement (often referred to as a nondisclosure agreement). A list of any such agreements in the résumé is unwarranted because it might be considered a form of egotistical disclosure and cause the attorney to turn away from that candidate. At the appropriate time, any governmental security clearance should be declared (not in the résumé) to the attorney during the initial telephone conversation or, at the latest, during the first face-to-face conversation.

3.8 CONCLUSIONS

Most of this discussion has focused on the scientist or engineer who may have been retained to be a designated expert witness. Remembering that one of the primary goals of the opposing legal counsel is to destroy the credibility of the expert witness, which can then nullify his or her testimony, for the scientist, engineer, and attorney (who will seek to retain a scientist or engineer as an expert witness), the following seven points are indicators of the capabilities of what it takes to be an expert witness:

- knowledge, training, skills, and experience in a field of specialty pertinent to the issues involved in the dispute;
- confidence in his or her expertise to research and develop opinions and conclusions about the subjects under dispute;
- time to research the issues and to develop an opinion and conclusions about the subjects under dispute;
- patience to wait weeks, months, or even a year or more before being called upon to testify;
- ability to speak clearly and convincingly to explain technical subjects in a simple manner before lawyers, judges, and juries;
- ability to remain calm and unemotional when testifying under oath; and

ability to smile courteously when in the witness chair, even to the most hostile opposing counsel.

Only when the scientist or engineer is comfortable with these points can he or she claim to be ready for the witness stand.

4 Attorney–Expert Witness Relationships

4.1 INTRODUCTION

The attorney–expert team is critical to the successful litigation of complex cases. But remember, there is always potential tension in this relationship; experts must be *objective,* but attorneys can lean toward being *subjective* insofar as attorneys must advocate their client's case. The lawyer assumes the role of zealously interpreting (but not creating) evidence in favor of his or her client. On the other hand, the expert witness must, with zeal that is at least equal to that of the attorney, assiduously search for facts and the truth.

The two—the attorney and would-be expert—often have not met each other before the initial contact. Generally, an attorney hears about or knows a scientist or engineer and then makes contact with that individual as a potential testifying expert. If the scientist or engineer does not know the attorney, it may be wise to check up on him or her through acquaintances or the comprehensive directory of opposing counsels; there is a listing for each state in the United States available in any law library.

The attorney–expert relationship starts after the attorney contacts the expert by means of a letter or, more commonly, by telephone. The first question that the attorney should ask the would-be expert relates to the scientist's or engineer's knowledge in the area of science or engineering under dispute or litigation and whether he or she will be available. On the other hand, the expert's curiosity may lie in the timetable of the litigation; it is most unwise to accept an assignment if the trial date is set within a few days because there will not be enough time for the expert to evaluate the case and perform the necessary literature review. At this time, the attorney will most likely ask the expert for a copy of his or her résumé and fee schedule.

Beyond this, once it has been determined that further conversations should occur—usually in the form of an interview—this is the time when the scientist or engineer will learn some, though perhaps not all, of the technical facts of the case and the role that the attorney expects the scientist or engineer to play as a witness in the dispute or litigation. This will, of course, be from the attorney's point of view, which should give pointers on being an effective consultant and witness.

4.2 CONTACT

The first step is for the attorney to decide if an expert is really needed for the case under consideration. If the decision is to retain an expert, contact must be made at once, and the attorney will request a copy of the scientist's or engineer's résumé and fee schedule. After the initial contact is made, the scientist or engineer should request a formal letter of intent (e-mail will suffice, although top law firms send a formal letter) stating that he or she is to be retained as a consultant or an expert as discussed in the telephone call or initial contact. The letter should have a brief outline of the duties, the dates when court activity is expected, and the agreed upon compensation. The letter should also have a statement regarding confidentiality of any information passed to the would-be expert.

By this time, the expert witness should have submitted to the attorney the following list of items:

- whether the scientist or engineer is hired as a consultant or an expert;
- the exact date the attorney expects the expert to start (which also may be requested by the judge or the opposing counsel);
- the scope of work to be done; and
- the fee schedule.

The scientist or engineer should not fall into the trap of talking to an attorney, doing some preliminary work, and then getting a letter some time later stating that the case has been settled and the service of the expert witness is not needed. Alternatively, the potential expert witness may hold the time available for the attorney but never hear from him or her again—even direct telephone calls to the attorney remain unanswered.

The expert often can offer the attorney information about various other experts whom the opposing counsel may use. Many scientists and engineers know each other and are often somewhat knowledgeable about each other's qualifications, good points, and "flaws." Sometimes the expert may even know the opinions of the opposing expert and suggest deposition questions for the attorney to ask.

Sometimes the expert needs to obtain help from another scientist. The expert scientist or engineer may need more information on the method of analysis of an agent or on a complex mixture; a chemist is then consulted. He or she may need more information on the side effects of long-term medication with painkilling drugs; a pharmacologist or pharmacist is then consulted. The attorney must give guidance to the expert on how this outside consultation should be described. In some states, the list of the expert's consultants need not be revealed in the written opinion; many courts assume that scientists talk to one another.

What the expert offers the attorney, in summary, is to educate the attorney on the science of the case and to fill in the gaps as necessary. The expert can help draft questions for the interrogatories to be presented to the opposing counsel as well as help the attorney answer questions given to him or her. An expert may suggest sources of information and evidence. Sometimes a consultant can help prepare another expert whom the attorney wishes to retain. Finally, one of the most important functions an expert performs is to give the attorney a scientific opinion based on the facts of the case.

4.2.1 Initial Work

The expert should investigate every aspect of the case; he or she must read all written material (that the attorney wishes to have revealed). The expert must evaluate the weaknesses as well as the strengths of the scientific case being prepared by the attorney. It is incumbent on the expert to inform the attorney what, if any, the defects of the case under consideration are; this should only be done orally unless otherwise requested. Usually, the attorney knows the strengths of the case. The expert must be on solid ground when this topic is discussed, and, if at all possible, the expert should have specific scientific references to share with the attorney on both the strengths and the weaknesses of the case.

For example, the scientist or engineer should determine from the attorney the general and specific objectives of the case. He or she must understand that the attorney can only give a limited amount of information, and there may be a number of facts that, for various reasons, are not disclosed to the scientist or engineer. Usually the monetary aspects of the case are not discussed. Because the expert should have no financial stake in the outcome of a case, the dollar values should not be of concern.

Any information that the scientist or engineer is given during this phase of the relationship with the attorney is subject to discovery and will most likely come out at the deposition and trial. A lack of certain details that are not instrumental in assisting the would-be expert to form an opinion is not a handicap and can protect the knowledge of the attorney. If applicable, the expert should learn from the attorney any previous legal precedents that pertain to the current trial. The attorney does not need to give references to specific court cases (unless it will help the expert), but the facts of the precedents can be useful.

The attorney should be specific about what is required in the way of communication from the expert to the attorney or from the expert to any other person the attorney designates. A scientist or engineer should not take it upon himself or herself to communicate to another person about the aspects of the case. In addition, the attorney must be specific as to how reports are to be made—in writing, by fax, by telephone, or face to face. The attorney should remind the expert about confidentiality and that every precaution must be taken not to give any confidential material to anyone, especially to the opposing group.

If the attorney fails to give the expert the information that he or she believes to be necessary, the expert should not hesitate to ask outright. It is also important for the scientist or engineer to remember that this is not an open forum for discussion of ideas and if the attorney does not satisfy the request for more knowledge or facts, it is because of the nature of the case and not necessarily a lack of trust.

During the early meetings, the scientist or engineer should talk with the attorney about the procedures that will be encountered in the courtroom. Even the seasoned expert witness can find changes in courtroom protocols from one state to the next or from state court to federal court.

At some time, the attorney will be expected to discuss attorney–expert privilege and work-product doctrines as well as when the expert's work may be covered by such privileges. Also, the would-be expert should be told how the privileges are waived under the rules of the state, province, or country where the dispute has been filed.

The safest approach for the would-be expert is to assume that, whatever is discussed, the expert must remember, at all times, that there is no attorney–client privilege in the relationship between expert and attorney. All written material, whether it is formal or simply notes, and possibly much conversation can be subject to discovery. Again, it is necessary to advise the expert not to write anything to the attorney without a specific request by the attorney to do so. This advice does not apply to the consultant or the nontestifying expert.

Only under unusual circumstances can there be an attorney–expert privilege. Both the attorney and the expert should conduct themselves as though there is no privilege because there may be no such protection. Opposing counsel can, by virtue of discovery, learn all information the expert provides to the attorney. However, specific situations can occur where privilege can be claimed. In some cases, the expert may look over very confidential documents that are labeled "classified" by the U.S. government; in such cases, the information may be considered privileged.

Briefly, *discovery* (Chapter 6) is a set of pretrial procedures, and there are rules relating to discovery (e.g., Rules of Civil Procedure, Rule 26; General Provisions Governing Disclosure; Duty of Disclosure). The attorney should advise the expert what is required, including limitations and exceptions. Discovery is one of the first phases of the pretrial; it is the means used by either party in a case to obtain facts and information about the case from the other party. This information is needed for the preparation for the trial. The sharing and exchanging of information may not be as complete as either side wishes without giving away information to the opposing side.

The rationale behind the discovery process is to permit the two parties to arrive at the truth, to prevent perjury, and to detect fraudulent claims and defenses. The discovery process educates both sides prior to the trial as to the true value of the claim. Discovery can facilitate settlement, speed up litigation, avoid surprises, and limit the issues. However, discovery is not simple; some attorneys may go beyond obtaining documents by delving into the private lives of the opponents and the experts.

Any communication between an attorney and a consultant who is not to be a trial witness is privileged. If, however, the consultant later becomes a testifying expert, there will be problems related to discovery; this is the reason it is essential to note the exact date that a consultant is designated an expert.

The items subject to discovery include all written reports of an expert (but not of a consultant or nontestifying expert). All documents (which may not be complete or comprehensive), depositions, interrogatories, and testimony, if available, are subject to discovery.

The attorney should explain to the expert that the implication is that anything written—personal notes, short communications to the attorney, or formal reports—is subject to discovery by opposing counsel. The attorney must give guidance to the expert on how to deal with audiotapes and videotapes. In this computer age, much information is stored on computer hard drives or on storage disks. As information becomes more complete, documents are altered and edited and previous versions become unavailable through overwriting. The expert should keep the attorney abreast of all files, thereby identifying all materials that are responsive to any request for information from an opposing counsel.

4.2.2 Experimental Work

When a scientist or engineer is retained as an expert, there is always the potential for experiments to be conducted in a laboratory. The need for experimental work or testing work should be discussed as soon as possible in the attorney–expert discussion because of the preparation time that may be necessary to initiate the work. The expert *must* be in a position to explain the mechanics of the chain of custody concept.

Questions then arise as to whether the experiment should be conducted by the expert in his or her laboratory or contracted out to a commercial laboratory. If the work is to be contracted out to a commercial laboratory, the attorney will most likely pass the obligation to the expert to make a thorough literature search to determine whether the proposed experiment has already been carried out by some other laboratory. When this is done, the expert may determine that experiments need to be conducted by an independent laboratory.

In either case (i.e., use of the expert's laboratory or an independent laboratory), the attorney should explain to the expert that he or she will be expected to monitor the work and that, no matter who does the work, complete records must be kept. Every observation, every calculation, and every measurement must be recorded. A bound (not loose-leaf) notebook is essential, and every page must be initialed and dated. Every purchase must be noted, including from whom, when, cost, purity (if a chemical), and amount obtained.

If the expert is unaware of this (and many scientists and engineers are), the attorney needs to explain that the concept of the *chain of custody* (*sample history*) is real in the court. Cases have been lost when the chain of custody was not complete.

4.2.2.1 Sample Acquisition

The value of any experimental work is judged not only by the characteristics of the work but also by the characteristics of the sample used for the laboratory tests. Many scientists and engineers do not appreciate this fact. Attorneys must, by default, also be included with the scientists and engineers as being reasonably unknowledgeable when it comes to experimental work that is beyond reproach.

The sample used for any such laboratory work must be representative of the bulk material; otherwise, data will be produced that are not representative of the material and that will, to be blunt, be incorrect no matter how accurate or precise the laboratory method is. In addition, the type and cleanliness of containers used for the starting materials and products are important. If the container is contaminated or is made of material that either reacts with the product or is a catalyst, the test results may be wrong. If the work involves instrumental investigation, the laboratory worker must be absolutely sure (i.e., beyond the shadow of a reasonable doubt) that the instruments are calibrated and more than adequate for the task.

Thus, the importance of the correct laboratory protocols (and these will vary with the type of laboratory investigation) destined for presentation in the courtroom should always be overemphasized. Incorrect procedure protocols can lead to erroneous analytical data from which performance of the product in service cannot be accurately deduced. Indeed, procedures that are perceived to have faults (even though they do not) because of incorrect presentation to the judge and jury will come under the angry glare and examination of opposing counsel (and his or her experts) and be dismissed without further ado. Detailed records of the circumstances and conditions during sampling have to be made, for example, in sampling from storage tanks; the temperatures and pressures of the separation plant and the atmospheric temperature would be noted.

For analytical work, representative samples are prerequisite for the laboratory evaluation of any type of product, and many precautions are required in obtaining and handling representative samples. The precautions depend upon the sampling procedure, the characteristics (low-boiling or high-boiling constituents) of the product being sampled, and the storage tank, container, or tank carrier from which the sample is obtained. In addition, the sample container must be clean, and the type to be used depends not only on the product but also on the data to be produced.

The basic objective of each procedure is to obtain a truly representative sample or, more often, a composite of several samples that can be considered to be a representative sample. In some cases, because of the size of the storage tank and the lack of suitable methods of agitation, several samples are taken from large storage tanks in such a manner that the samples represent the properties of the bulk material from different locations in the tank; thus, the composite sample will be representative of the entire lot being sampled. This procedure allows for differences in samples that might be due to the stratification of the bilk material due to tank size or temperature at the different levels of the storage tank.

Solid samples require a different protocol that might involve melting (liquefying) the bulk material (assuming that thermal decomposition is not induced), followed by homogenization. On the other hand, the protocol used for coal sampling might also be applied to sampling petroleum products, such as coke, that are solid and for which accurate analysis is required before sales.

4.2.2.2 Chain of Custody

Once the sampling procedure is accomplished, the sample container should be labeled immediately to indicate the product, time of sampling, location of the sampling point, and any other information necessary for the sample identification. If the samples were taken from different levels of the storage tank, the levels from which the samples were taken and the amounts taken and mixed into the composite should be indicated on the sample documentation.

Although the preceding text focused on the acquisitions of samples from various storage sources, the *chain of custody* (*sampling records*) for any procedure must be complete and should include, but is not restricted to, information such as:

1. the precise (geographic or other) location (or site, refinery, or process unit) from which the sample was obtained;
2. the identification of the location (or site, refinery, or process unit) by name;
3. the character of the bulk material (solid, liquid, or gas) at the time of sampling;
4. the means by which the sample was obtained;
5. the means and protocols that were used to obtain the sample;
6. the date and the amount of sample that was originally placed into storage;
7. any chemical analyses (elemental analyses, fractionation by adsorbents or by liquids, functional type analyses) that have been determined to date;
8. any physical analyses (API gravity, viscosity, distillation profile) that have been determined to date;
9. the date of any such analyses included in items 5 and 6;
10. the methods used for analyses that were employed in steps 5 and 6;
11. the analysts who carried out the work in steps 5 and 6; and
12. a log sheet showing the names of the persons (with the date and the reason for the removal of an aliquot) who removed the samples from storage and the amount of each sample (aliquot) that was removed for testing.

In summary, there must be a means to identify the sample history as carefully as possible so that each sample is tracked and defined in terms of source and activity. The perception that the chain of custody has been broken and the sample under examination has been compromised leaves the data open to (usually) successful attack (by opposing counsel) as well as an attack on the reliability of the data (also by opposing counsel); the expert scientist or engineer is left with … nothing. However, in another way, he or she may be left with something: (1) the ire of the court, (2) the disbelief of the jurors, and (3) the ire of the attorney.

Thus, the perception of the accuracy of the data from the laboratory procedures and tests for which the sample is used has, through a faulty *chain of custody*, placed what might have been valuable evidence into the category commonly known as *beyond a reasonable doubt* and the case is lost.

4.2.2.3 Measurement

The issues that face the scientist and engineer in the laboratory analysis include the need to provide higher-quality results. In addition, environmental regulations may influence the method of choice. Nevertheless, the method of choice still depends, to a large extent, on the characteristics of the sample to be analyzed and the accuracy of the experimental method.

The *accuracy* of a laboratory procedure is a measure of how close the experimental result will be to the true value of the reaction or property being measured. As such, the accuracy can be expressed as the *bias* between the test result and the true value. However, the *absolute accuracy* can only be established if the true value is known from other work.

In the simplest sense, a convenient method to determine a relationship between two measured product yields or product properties is to plot one against the other. Such an exercise will provide either a line fit of the points or a spread that may or may not be within the limits of experimental error. The data can then be used to determine the approximate accuracy of one or more points employed in the plot.

The *precision* of a test method is the variability between test results obtained on the same material, using a specific test method or laboratory procedure. The precision of a test or procedure is usually unrelated to its accuracy. The results may be precise but not necessarily accurate. In fact, the precision of a method is the amount of scatter in the results obtained from multiple analyses of a homogeneous sample. To be meaningful, the precision study must be performed using the exact reactants or sample and standard preparation procedures that will be used in the final method. Precision is expressed as repeatability and reproducibility.

For some cases, the court may insist on a laboratory procedure's being performed not only by the scientific or engineering expert but also by a third party from within the same laboratory or even from a completely different laboratory. This brings into operation the concept of intralaboratory (within-laboratory) precision or interlaboratory precision. The former refers to the precision of a procedure or test method when the results are obtained by the same operator in the same laboratory using the same apparatus. The latter refers to the precision of a procedure or test method when the results are obtained by a different experimentalist or operator in a different laboratory using the same procedure or test method. In some cases, the precision is applied to data gathered by a different operator in the same laboratory using the same apparatus. Thus, intralaboratory precision has an expanded meaning insofar as it can be applied to laboratory precision.

The repeatability or repeatability interval of a procedure or test method (r) is the maximum permissible difference due to test error between two results obtained on the same material in the same laboratory:

$$r = 2.77 \times \text{standard deviation of procedure or test}$$

The repeatability interval (r) is, statistically, the ninety-five percent probability level; that is the differences between two test results are unlikely to exceed this repeatability interval more than five times in a hundred.

The interlaboratory precision (or between-laboratory precision) is defined in terms of the variability between test results obtained on the aliquots of the same homogeneous material in different laboratories using the same test method.

The term *reproducibility* or *reproducibility interval* (*R*) is analogous to the term *repeatability,* but it is the maximum permissible difference between two results obtained on the same material but now in different laboratories. Therefore, differences between two or more laboratories should not exceed the reproducibility interval more than five times in a hundred:

$$R = 2.77 \times \text{standard deviation of test}$$

The repeatability value and the reproducibility value have important implications for quality. As the demand for clear product specifications and hence control over product consistency grows, it is meaningless to establish product specifications that are more restrictive than the reproducibility/repeatability values of the specification test methods.

Once the reliability of the data has been ensured, opposing counsel will investigate one other facet if he or she has experts who understand the various aspects of laboratory investigations and laboratory testing: method validation.

4.2.2.4 Method Validation

Method validation is the process of proving that an experimental procedure or analytical method is acceptable for its intended purpose. Many organizations, such as the American Society for Testing and Materials (ASTM), provide a framework for performing such validations. In general, methods for experimental reliability, product specifications, and regulatory submission must include studies on specificity, linearity, accuracy, precision, range, detection limit, and quantitation limit.

The process of method development and validation covers all aspects of the analytical procedure, and the best way to minimize method problems is to perform validation experiments during development. In order to perform validation studies, the approach should be viewed with the understanding that validation requirements are continually changing and vary widely, depending on the type of product under test and compliance with any necessary regulatory group.

In the early stages of new product development, it may not be necessary to perform all of the various validation studies. However, the process of validating a method cannot be separated from the actual development of the method conditions because the developer will not know whether the method conditions are acceptable until validation studies are performed. The development and validation of a new analytical method may therefore be an iterative process. Results of validation studies may indicate that a change in the procedure is necessary, which may then require revalidation. During each validation study, key method parameters are determined and then used for all subsequent validation steps.

The first step in the method development and validation cycle should be to set minimum requirements, which are essentially acceptance specifications for the method. A complete list of criteria should be agreed on with end users during method development before the method is developed so that expectations are clear. Once the validation studies are complete, the method developers should be confident in the ability of the method to provide good quantitation in their own laboratories. The remaining studies should provide greater assurance that the method will work well in other laboratories, where different operators, instruments, and reagents are involved and where it will be used over much longer periods of time.

The remaining precision studies constitute much of what is often referred to as *ruggedness. Intermediate precision* is the precision obtained when an assay is performed by multiple analysts using several instruments on different days in one laboratory. Intermediate precision results are used to identify which of the previous factors contributes significant variability to the final result.

The last type of precision study is *reproducibility* (q.v.), which is determined by testing homogeneous samples in multiple laboratories, often as part of interlaboratory crossover studies. The evaluation of reproducibility results often focuses more on measuring bias in results than on determining differences in precision alone. Statistical equivalence is often used as a measure of acceptable interlaboratory results. An alternative, more practical approach is the use of *analytical equivalence,* in which a range of acceptable results is chosen prior to the study and used to judge the acceptability of the results obtained from the different laboratories.

Performing a thorough method validation can be a tedious process, but the quality of data generated with the method is directly linked to the quality of this process. Time constraints often do not allow for sufficient method validation. Many researchers have experienced the consequences of invalid methods and realized that the amount of time and number of resources required to solve problems discovered later exceed what would have been expended initially if the validation studies had been performed properly. Putting in time and effort up front will help any analyst to find a way through the method validation maze and eliminate many of the problems common to analytical methods that have been inadequately validated.

4.2.2.5 Other Information

The function of the expert is to provide the expertise on the science of the case, and the function of the attorney is to provide the expertise on the law; however, the two are not mutually exclusive. The attorney may wish to become knowledgeable about the science of the case. The various experts may be asked to present seminars for the attorney and also for interested associates and paralegals on the case. Indeed, if the expert knows some legal precedents, this information should be disclosed during one of the many discussions with the attorney. Such an expert may even *suggest* (not *dictate*) some litigation strategy to the attorney.

The attorney should explain any important laws and decisions that will pertain to the case to the expert, and both should review the function of an expert witness in accordance with the *Frye* rule and the *Daubert* decision (Chapter 1), plus any other recent decisions of the Supreme Court. A clear understanding of the rules will enable the scientist or engineer to be a better witness. Also, it will help him or her to evaluate the credentials of the expert witnesses retained by the opposing counsel.

The expert must review the facts of the case in great detail, and a thorough literature search is normally in order. Many scientific and engineering databases are available, and many libraries now are online and have ready access to these databases. The search strategy must be carefully thought out to yield pertinent information.

Many personal meetings between the expert and the attorney are absolutely necessary. It is in these meetings that the expert should give the attorney the benefit of his or her evaluation of the case, which should be done verbally. The expert is advised, however, to keep a written log of time spent (and what was done) to formulate the opinion.

Only after a request should the expert tell the attorney the opinion or conclusion he or she has derived from the information available. A subtle point here is that the expert has not been retained by the attorney to give an opinion but, rather, to derive an opinion. The expert must be assured and be careful that the attorney is not trying to buy an opinion. The statement of the opinion should be clear and concise and understandable by a nontechnical audience—the judge and the jurors.

In short, the expert should investigate every aspect of the case. He or she must read all of the written material that the attorney wishes to have revealed. In addition, he or she must evaluate the weaknesses as well as the strengths of the scientific case being prepared by the opposing counsel. In fact, it is incumbent on the expert to inform the attorney what, if any, the defects of the case under consideration are; this should only be done orally unless otherwise requested. Usually, the attorney knows the strengths of the case and the expert must be on solid ground when this topic is discussed. If at all possible, the expert should have specific scientific references to share with the attorney on both the strengths and the weaknesses of the case.

Finally, the expert often can offer the attorney information about the various other experts whom the opposing counsel may have retained. Many scientists and engineers know each other and are often somewhat knowledgeable about each other's qualifications, good points, and flaws. For example, the other expert may have a tendency to ramble, or his or her ego (as the top authority on the technical issues under dispute) will play into the hands of a skillful cross-examiner and offer evidence in favor of the other side. Sometimes the expert may even know the opinions of the opposing scientific or engineering expert and be able to suggest particularly searching and revealing questions during the opposing expert's deposition.

4.3 SCIENTISTS AND ENGINEERS

The attorney may have little experience of scientists and engineers, so it is necessary that he or she discover the scientific or engineering psyche at the earliest possible time. For example, the attorney may engage a scientist who always, by virtue of training, has twenty more experiments to do before he or she will give the attorney a "possible maybe." On the other hand, the attorney may engage an engineer who believes that he or she is the most logical person since the Vulcan Mr. Spock and that whatever he or she writes on paper, whatever the assumptions, is always correct.

At a very early stage of the interview process, the attorney must weed out such egotistical candidates and move on to the next possible believable and less egotistical candidate. More specifically, the attorney is seeking the scientist or engineer who has special knowledge or skill gained by education, training, or experience; he or she may be retained by an attorney and then summoned to court and designated as an expert (based on that person's field of expertise) to give expert evidence during a trial from which opinion and conclusions follow. Pretrial preparation by the expert witness refreshes the level of expertise, enhances the quality of the opinion expressed, reduces stress, and saves time.

The rules governing the admissibility of expert testimony are the domain of the lawyer and the trial judge. It is unnecessary for the expert witness to be familiar with the intricacies and nuances of expert testimony and its frequent partner, hearsay evidence. The admissibility of expert evidence is predicated on the existence of knowledge and experience that is beyond that of the ordinary citizen and is applicable to the matter before the court. Such evidence must have the effect of proving facts.

Engineers and scientists are often confused in the minds of the general public, with the former being closer to applied science. Scientists explore nature in order to discover general principles; engineers apply established principles drawn from mathematics and science in order to develop economical solutions to technical problems. In short, scientists study things, whereas engineers build things. But there are plenty of instances where significant accomplishments are made in both fields by the same individual. Scientists often perform engineering tasks in designing experimental equipment and building prototypes, and some engineers do first-rate scientific research. Chemical, electrical, and mechanical engineers are often at the forefront of scientific investigation of new phenomena and materials.

4.3.1 Science in the Court

In the broadest sense, *scientist* refers to any person who engages in a systematic activity to acquire knowledge or an individual who engages in such practices and traditions that are linked to schools of thought or philosophy. In a more restricted sense, *scientist* refers to individuals who use the scientific method. The person may be an expert in one or more areas of science.

Historically, scientists were termed natural philosophers or men of science because they were men of knowledge. *Science* and *philosophy* were roughly synonymous. The term *scientist* was coined in 1833 to describe an expert in the study of nature, but this term did not gain wide acceptance until the turn of the nineteenth century. By the twentieth century, the modern notion of science as a special brand of information about the world practiced by a distinct group and pursued through a unique method was essentially in place.

Science and technology have continually modified human existence. As a profession, the scientist of today is widely recognized. Scientists include theoreticians, who mainly develop new models to explain existing data, and experimentalists, who mainly test models by making measurements; in practice, the division between these activities is not clear-cut, and many scientists perform both.

Scientists can be motivated in several ways. Many have a desire to understand why the world is as we see it and how it came to be. They exhibit a strong curiosity about reality. Other motivations are recognition by their peers, prestige, and the desire to apply scientific knowledge for the benefit of people's health, nations, the world, nature, or industries. Only a few scientists count generating personal wealth as an important driving force behind their science.

The philosophy of science that a court draws so heavily upon focuses on the nature of scientific investigation and informs virtually all of modern scientific inquiry. The philosophy of science provides the framework that practitioners in all of these disciplines use to analyze data to find out whether their theories are correct. Once one understands the philosophical basis of science upon which a court relies, much of the statistical part of scientific testimony makes sense; this represents half—and perhaps more—of understanding the entirety of the expert testimony that is offered in courts at the present.

When the U.S. Supreme Court handed down its opinion in *Daubert v. Merrell Dow Pharmaceuticals, Inc.* (Chapter 1), it began a wide-ranging debate about the rules that govern the admissibility of expert testimony in both state and federal trials. Many articles have been published in response to the decision, and there are several *Daubert* Web sites. This vigorous response is not surprising, because *Daubert* held that the Federal Rules of Evidence had displaced the fifty-year-old *Frye* "generally accepted" standard for the admissibility of scientific testimony in federal trials and then determined a new standard for admitting expert scientific testimony in a federal trial.

Despite substantial disagreement in the legal community about what *Daubert* really means, the meaning of a court's scientific dialogue is fairly clear. Even though the scientific principles that a court articulates are ultimately discussed as scientific and statistical concepts that are somewhat alien to the legal system, the basis of scientific evidence is found in the principles of philosophy and logic that have long informed the legal system.

4.3.2 Engineering in the Court

An engineer is someone who is trained or professionally engaged in a branch of engineering. Engineers use technology, mathematics, and scientific knowledge to solve practical problems. People who work as engineers typically have an academic degree (or equivalent work experience) in one of the engineering disciplines.

Engineers consider many factors when developing a new product. For example, in developing an industrial robot, engineers precisely specify the functional requirements; design and test the robot's components; integrate the components to produce the final design; and evaluate the design's overall effectiveness, cost, reliability, and safety. This process applies to the development of many different products, such as chemicals, computers, engines, aircraft, and toys.

In addition to design and development, many engineers work in testing, production, or maintenance. These engineers supervise production in factories, determine the causes of component failure, and test manufactured products to maintain quality. They also estimate the time and cost to complete projects. Some move into engineering management or sales. In sales, an engineering background enables them to discuss technical aspects and assist in product planning, installation, and use. Supervisory engineers are responsible for major components or entire projects.

In the United States, engineering degrees range from a bachelor's degree in sciences or engineering (four years) to a master's degree in sciences or engineering (adding one or two years, depending on the university) to a doctor of engineering, which entails completing original research.

4.3.3 The Scientific Method and the *Daubert* Rule

4.3.3.1 Hypothesis Testing

Hypothesis testing is the process of deriving a hypothesis (or proposition) about an observable group of events from accepted scientific principles and then investigating whether, upon observation of data regarding that group of events, the hypothesis seems true. Hypothesis testing distinguishes the scientific method of inquiry from nonscientific methods, and the scientific method of inquiry is required for the resulting inferences to be the basis of admissible expert testimony. Examples of hypothesis testing, and therefore the scientific method alluded to in the *Daubert* decision, are plentiful in all of the most highly regarded professional journals that publish empirical research.

4.3.3.2 The Known or Potential Error Rate

The second parameter that *Daubert* suggests that a trial judge use in evaluating the scientific validity and therefore evidentiary reliability of purported scientific testimony is the *known or potential rate of error* associated with using the particular scientific technique. Simply, this is the likelihood of being wrong that the scientist associates with the assertion that an alleged cause has a particular effect. Most scientists routinely require that this error rate be very small, usually between one and five percent. Errors in the laboratory (often called *experimental differences,* which sounds better than *errors*) are preferred to be within three percent (either side) of the real number.

There are two types of error rates in testing hypotheses: *type I error* and *type II error.* Type I error is the procedure's or test's propensity for *false positives,* and type II error is the test's propensity for *false negatives.* The type I error is the most commonly cited component of the error rate in hypothesis testing and is also known as the *level of confidence* of the hypothesis test and as the level of statistical significance of the test's result. Determining this error rate is actually part of conducting a hypothesis test and relates to accuracy and precision.

The relationship between the first two criteria, the hypothesis test and the error rate, is so close that it is virtually unheard of for a scientist to report that a hypothesis was rejected without stating the level of confidence at which it was rejected. Such a report would be completely meaningless.

The third criterion that the Supreme Court suggested for use by trial courts in determining whether expert testimony reaches the trier of fact is *whether the theory or technique has been subjected to peer review and publication.*

Publication is typically the purpose for which research is offered up for peer review, and passing the peer review is required for publication. Peer review and publication of a scientist's work are largely a term of art that means that the scientist's peers have sanctioned the work as credible and accepted it for publication. Publication then exposes the work to further review by other scientists, whose responses to the research indicate their agreement or disagreement with the methods and results of the work. Scientists' peers often express agreement with the work of a particular scientist by citing the work with approval or as authority, or by extending the work. Properly executed hypothesis tests with their attendant error rates are the essence of the scientific method and are very nearly necessary conditions for peer review to result in publication.

The fallacy behind peer review and publication is that many reviewers do not recognize any work that disagrees with their own. There are those few reviewers who will recognize disagreement and a scientist's or engineer's right to publish as long as the work is well thought out and presented logically and without errors.

General acceptance is a summary measure of the extent to which the expert's methods produce information that qualifies as scientific knowledge. Scientific methods begin the process of becoming generally accepted in the scientific community by bringing appropriate hypothesis testing techniques to bear on questions (or hypotheses) of interest to the scientific community in a fashion that results in the peer approval required for publication. They move toward general acceptance by then withstanding the scrutiny of the broader scientific community to which publication exposes the methods. The inescapable conclusion must be that the relevant scientific community within which the technique finds acceptance must be the community of real-world scientists who pursue science for non-litigation purposes and that finding general acceptance within the community of forensic scientists does not constitute general acceptance in the relevant scientific community.

However there is always the possibility (hopefully, extremely remote) that the experts retained by opposing counsel may be "hired guns" who will propose a sham technique that serves their collective purpose. It is equally possible that a plaintiff's hired experts could propose another sham technique that serves their purposes. Both techniques would be supported for admissibility by the general acceptance criteria, despite the fact that they were both sham techniques.

It is interesting to note that a scientist reading *Frye* and *Daubert* might say that *Daubert* explains *Frye* at least as much as it displaces it. *Frye* defined the evidentiary issue as reliability and then deferred fully to an amorphously defined scientific community for its general acceptance, which it used for evidentiary reliability.

The *Daubert* rule still looks to the scientific community for its general acceptance as an indicator of evidentiary reliability, but it goes further and defines that general acceptance in two ways. First, it addresses the characteristics of the scientific community to which it defers for the general acceptance that it uses as a determinant of evidentiary reliability. Second, the *Daubert* rule recognizes that there is a basic structure of inquiry known as the scientific method that is the standard used across different branches of science for their scientific investigations, and it requires that the science proffered to the federal bench be grounded in that basic structure. This requires posing and testing hypotheses and specifying the rates of error for those hypothesis tests.

4.4 INFORMATION FROM THE ATTORNEY

There are several aspects of the potential work that the expert should discuss with the attorney at an early stage of the contact. Even though the answers to some of these issues may not be known at the time, it is good to lay them on the table for future consideration. Consideration of these points should help the scientist or engineer to reduce the impact of a court summons on a professional's ability to carry out his or her primary goal.

The most important aspect of being retained as an expert is to clarify precisely what area of expertise is expected. There should be an understanding of the expertise required as soon as possible, because this will focus preparation if the request is written and reduce misunderstanding later.

At this time, the expert should also request the names and areas of expertise of any other experts that the attorney intends to use. At some time, as the dispute evolves, a list of experts retained by the opposing counsel will also become available. Because expert witnesses do not always agree on the interpretation or effects of specific circumstances or facts, knowing the names of other experts will help the expert formulate responses to the challenges of his or her opinion through other expert witnesses. If the witness has some understanding of a challenge, it may enhance the quality of his or her evidence as well as what it may entail to prepare to respond.

The expert also needs to clarify as soon as possible the extent of his or her pretrial involvement in the forms of meetings with counsel as well as attendance at or formulation of questions for pretrial proceedings such as discoveries.

Written reports form the basis for pretrial preparation, settlement negotiations, and testimony during trial. The expert should determine any due date for the report and to whom it is directed. The reports may lead to a decision not to call the expert witness or lead to a settlement and hence prevent a trial. A written report should incorporate only what is necessary. To prevent misunderstandings, the expert should request from the attorney a list of questions or an outline of the issues the report should address. As a minimum, the report should identify the reason for the report, the matter to be explored, the witness's observations and rationale, and other significant information and sources. In addition, the report should state any conclusions the witness may have reached based on the observations and information.

Usually, the expert is summoned to appear at a deposition or at trial though issuance of a subpoena. Discussion with the attorney at the very beginning of the contact can remove some of the hesitancy and stress associated with receiving a subpoena. After that, it is a matter of following the instructions contained in the order. If the summons is received through the attorney, he or she will inform the expert of this. If the summons is received directly from the court, the expert should notify the attorney. The attorney will advise the expert to take only what the subpoena requires. Any notes or documents that are used to answer an inquiry are subject to retention by opposing counsel for further study. The document or notes may then become the subject of further examination by opposing counsel or the court.

At some time early in the attorney–expert contact, a discussion of the time required to attend the trial is necessary. Trials requiring expert testimony may be time consuming. Many expert witnesses have other duties and, although it may not be granted, the witness may request a convenient time to testify. The attorney can also estimate the length of time that the witness may expect to be at the court.

The expert will also need to address with the attorney the extent of any legal protection for testimony. Because professional persons function under codes of ethics and confidentiality, the expert should insist on clear understandings of what protection the court may provide and how it is provided for potential breaches of the code of ethics that may arise during testimony. Written confirmation from counsel outlining the protection should be obtained before testifying.

In the early stages of the attorney–expert contact, the attorney should explain that the expert should be objective and base his or her opinions and interpretations on sound professional knowledge. The quality and hence weight given to the witness's testimony will depend on credibility. Remember that the court requires interpretation and understanding of professional opinion. In addition, the expert should be instructed to answer questions in plain, understandable language. The expert witness attends court to interpret and express opinions about facts, and plain, understandable language will aid the court in understanding interpretations and opinions. The use of scientific or engineering jargon may lead to further questioning, resulting in confusion and perhaps a loss of credibility.

The expert should answer only what counsel or the court asks. If clarification or interpretation is needed, the expert can do so as necessary. It is better to acknowledge lack of expertise in a specific area than to risk misleading responses. Failure to acknowledge a possible second interpretation may result in a loss of credibility. The worst mistake that an expert can make is to assume that the judge and jurors are familiar with the profession, its descriptions, and its terminology.

The expert must assume that the evidence and the manner in which it is presented will be assessed for validity and weighed against other evidence.

The main objections raised by the opposing counsel to expert testimony arise because of hearsay and whether or not there is a solid technical basis for the opinions expressed. Opposing counsel may object to the question asked or to the answer given, and an objection will be raised. After an objection is raised, the court rules on the objection and instructs counsel how to proceed; the expert witness should refrain from speaking until the court instructs otherwise. The expert witness should not interrupt counsel or the judge as a means of quickly explaining away the cause for the objection; he or she should not attempt to justify comments unless asked to do so.

After the expert gives testimony, the judge permits him or her to leave the witness stand. The witness may be free to stay in the court or leave. Some judges prefer that expert witnesses appear in court only for testimony, do not appear too early (be called by a bailiff), and leave as soon as testimony is completed. The witness may, however, be required to remain for further testimony or to return as the judge instructs.

4.5 DISCOVERABLE AND NONDISCOVERABLE COMMUNICATIONS

Discovery is a formal investigation governed by court rules that is conducted before trial. Discovery allows one party to question other parties and, sometimes, witnesses. It also allows one party to force the others to produce requested documents or other physical evidence. The most common types of discovery are interrogatories, consisting of written questions the other party must answer under penalty of perjury, and depositions, which involve an in-person session at which one party to a lawsuit has the opportunity to ask oral questions of the other party or witnesses under oath while a written transcript is made by a court reporter.

Other types of pretrial discovery consist of written requests to produce documents and requests for admission, by which one party asks the other to admit or deny key facts in the case. One major purpose of discovery is to assess the strength or weakness of an opponent's case, with the idea of opening settlement talks. Another is to gather information to use at trial. Discovery is also present in criminal cases, in which, by law, the prosecutor must turn over to the defense any witness statements and any evidence that might tend to exonerate the defendant. Depending on the rules of the court, the defendant may also be obliged to share evidence with the prosecutor.

Discovery devices used in civil lawsuits are derived from the practice Rules of Equity, which gave a party the right to compel an adverse party to disclose material facts and documents that established a cause of action. The Federal Rules of Civil Procedure have supplanted the traditional equity rules by regulating discovery in federal court proceedings. State laws governing the procedure for civil lawsuits, many of which are based upon the federal rules, have also replaced the equity practices.

Discovery is generally obtained either by the service of an adverse party with a notice to examine prepared by the applicant's attorney or by a court order pursuant to statutory provisions. Discovery devices narrow the issues of a lawsuit, obtain evidence not readily accessible to the applicant for use at trial, and ascertain the existence of information that might be introduced as evidence at trial. Public policy considers it desirable to give litigants access to all material facts not protected by privilege to facilitate the fair and speedy administration of justice. Discovery procedures promote the settlement of a lawsuit prior to trial by providing the parties with opportunities to evaluate the facts before them realistically.

Discovery is contingent upon a party's reasonable belief that he or she has a good cause of action or defense. A court will deny discovery if the party is using it as a fishing expedition to ascertain information for the purpose of starting an action or developing a defense. A court is responsible for protecting against the unreasonable investigation into a party's affairs and must deny discovery if it is intended to annoy, embarrass, oppress, or injure the parties or the witnesses who will be subject to it. A court will stop discovery when it is used in bad faith.

Pretrial discovery is used for the disclosure of the identities of persons who know facts relevant to the commencement of an action but not for the disclosure of the identities of additional parties to the case. In a few jurisdictions, however, the identity of the proper party to sue can be obtained through discovery. Discovery pursuant to state and federal procedural rules may require a party to reveal the names and addresses of witnesses to be used in the development of the case.

Discovery is not automatically denied if an applicant already knows the matters for which he or she is seeking discovery, because one of its purposes is to frame a *pleading* in a lawsuit. On the other hand, discovery is permitted only when the desired information is material to the preparation of the applicant's case or defense. Discovery is denied if the matter is irrelevant or if it comes within the protection of a privilege.

Privileged matters are not a proper subject for discovery. For example, a person cannot be forced to disclose confidential communications regarding matters that come within the attorney–client privilege. Discovery cannot be used to compel a person to reveal information that would violate his or her constitutional guarantee against self-incrimination. However, if a party or witness has been granted immunity regarding the matters that are the basis of the asserted privilege, he or she can be required to disclose such information upon pretrial examination.

A person who refuses to comply with discovery on the basis of an asserted privilege must claim the privilege for each particular question at the time of the pretrial examination. An attorney or the court itself cannot claim the privilege for that person. However, a person may waive the privilege and answer the questions put to him or her during discovery.

A party may challenge the validity of a pretrial examination if asserted prior to trial. The merits of such an objection will be evaluated by the court during the trial when it rules on the admissibility of the evidence. If the questions to be asked during a discovery, such as the identity and location of a particular witness, pose a threat to anyone's life or safety, a party can make a motion to a court for a protective order to deny discovery of such information.

Failing to appear or answer questions at an examination before trial might result in a contempt citation, particularly if the person has disobeyed the command of a subpoena to attend. If discovery is pursuant to a court order, the court will require that the party's refusal to answer questions be treated as if the party admitted them in favor of the requesting party. Such an order is called a preclusion order because the uncooperative party is precluded from denying or contradicting the matters admitted due to his or her intentional failure to comply with a discovery order.

Only facts, not matters or conclusions of law or opinions, can be admitted when there is no disagreement between the parties. The requesting party does not have to make a motion before a court prior to making such a demand but must comply with any statutory requirements. The matters or documents to be admitted must be particularly described and there must be a time limit for a reply. The response should admit or deny the request or explain in detail the reason for refusing to do so. Failure to make a response within the specified time results in the matter being admitted, precluding the noncomplying party from challenging its admission during the trial.

4.6 INTERROGATORIES

Interrogatories are a formal set of written questions propounded by one litigant and required to be answered by an adversary in order to clarify matters of evidence and help to determine in advance what facts will be presented at any trial in the case. The forms that interrogatories assume are as various as the minds of the persons who propound them. They should be as distinct as possible and capable of a definite answer; they should leave no loopholes to an unwilling witness for evasion. Care must be observed to put no leading questions in original interrogatories, for these always lead to inconvenience; for scandal or impertinence, interrogatories will, under certain circumstances, be suppressed. The vast majority of such questions are to find background information about the litigants that is not specific to each case, so it is common to use preprinted forms containing standard questions that are generally relevant to the type of case at hand; these are called form interrogatories. They may even be determined by statute or court rules.

Thus, interrogatories are a discovery device used by a party, usually a defendant, to enable the individual to learn the facts that are the basis for or support a suit with which he or she has been served by the opposing party. They are used primarily to determine what issues are present in a case and how to frame a responsive pleading or a deposition. Only parties to an action must respond to interrogatories, unlike depositions, which question both parties and witnesses.

Interrogatories are used to obtain relevant information that a party has regarding a case, but they cannot be used to elicit privileged communications.

The question must be stated precisely to evoke an answer relevant to the litigated issues. A party can seek information that is within the personal knowledge of the other or that might necessitate a review of his or her records in order to answer.

When interrogatories seek disclosure of information contained in corporate records, the party upon whom the request is served can designate the records that contain the answers, thereby making the requesting party find the answer for himself or herself. No party can be compelled to answer interrogatories that involve matters beyond the party's control. Objections to questions submitted can be raised, and a party need not answer them until a court determines their validity. Interrogatories are one of the most commonly used methods of discovery. They can be employed at any time, and there is no limit on the number that can be served.

Normal practice is for the lawyers to prepare the questions and for the answering party to have help from its attorney in understanding the meaning (sometimes hidden) of the questions and to avoid any wording in answers that could be interpreted against the party answering. Objections as to relevancy or clarity may be raised either at the time the interrogatories are answered or when they are used in trial. Most states limit the number of interrogatories that may be asked without the court's permission to keep the questions from being a means of oppression rather than just a source of information. Although useful in getting basic information, they are much easier to ask than answer and are often intentionally burdensome. In addition, the parties may request depositions (pretrial questioning in front of a court reporter) or send requests for admissions, which must be answered in writing.

4.7 DEPOSITIONS

Deposition testimony is taken orally, with an attorney asking questions and the deponent (the individual being questioned) answering while a court reporter or tape recorder (or sometimes both) records the testimony. Deposition testimony is generally taken under oath, and the court reporter and the deponent often sign affidavits attesting to the accuracy of the subsequently printed transcript. Depositions are a discovery tool. Discovery is the process of assembling the testimonial and documentary evidence in a case before trial. Other forms of discovery include interrogatories (written questions that are provided to a party and require written answers) and requests for production of documents.

Depositions are commonly used in civil litigation (suits for money damages or equitable relief); they are not commonly used in criminal proceedings (actions by a government entity seeking fines or imprisonment). Some U.S. states provide for depositions in criminal matters under special circumstances, such as to compel statements from an uncooperative witness, and a few provide for depositions in criminal matters generally.

Before a deposition takes place, the deponent must be given adequate notice as to its time and place. Five days' notice is usually sufficient, but local rules may vary. Persons who are witnesses but not parties to the lawsuit must also be served with a subpoena (a command to appear and give testimony, backed by the authority of the court).

Depositions commonly take place after the exchange of interrogatories and requests for production of documents because the evidence obtained from the latter often provides a foundation for the questions posed to the deponent. Any documents, photographs, or other evidence referred to during the deposition is marked and numbered as exhibits for the deposition, and the court reporter attaches copies of these exhibits to the subsequent deposition transcript. Generally, at the outset of the deposition, the court reporter, who is often also a notary public, leads the deponent through an oath that the testimony that will be given will be true and correct.

The examining attorney begins the deposition and may ask the deponent a wide variety of questions. Questions that could not be asked of a witness in court because of doubts about their relevance or concerns about hearsay (statements of a third party) are usually allowed in the deposition setting because they might reasonably lead to admissible statements or evidence. A party who refuses to answer a reasonable question can be subject to a court order and sanctions. However, a party may refuse to answer questions on the basis of privilege (a legal right not to testify). For example, statements made to an attorney, psychiatrist, or physician by a client seeking professional services can remain confidential, and a client may assert a privilege against being required to disclose these statements.

After the examining attorney's questions are completed, the attorney representing the adverse party in the litigation is permitted to ask follow-up questions to clarify or emphasize the deponent's testimony. In litigation involving a number of represented parties, any other attorney present may also ask questions.

The court reporter often records the proceedings in a deposition on a stenographic machine, which creates a phonetic and coded paper record as the parties speak. Occasionally, an attorney or witness may ask the court reporter to read back a portion of previous testimony during the deposition. Most modern stenographic machines also write a text file directly to a computer diskette during the deposition. In the past, arduous manual labor was required to turn the phonetic and coded paper copy into a complete, hand-typed transcript. This is now rarely necessary because sophisticated computer programs can create a transcript automatically from the text file on the diskette. When the transcription is complete, copies are provided to the attorneys, and the deponent is given the opportunity to review the testimony and correct any typographic errors.

Because it is taken with counsel present and under oath, the deposition becomes a significant evidentiary document. Based upon the deposition testimony, motions for *summary judgment* or *partial summary judgment* as to some claims in the lawsuit may be brought. Summary judgment allows a judge to find that one party to the lawsuit prevails, without trial, if there are no disputed material facts, and judgment must be rendered as a matter of law. If motions for summary judgment are denied and the case goes to trial, the deposition can be used to impeach (challenge) a party or witness who gives contradictory testimony on the witness stand.

The advent of sophisticated and low-cost video technology has resulted in increased videotaping of depositions. Both sides must agree to the videotaping through a signed agreement called a stipulation, and in some jurisdictions, the parties must also seek a court order. A videotaped record of a deposition offers several advantages. First, a videotape shows clearly the facial expressions and postures of the witnesses, which can clarify otherwise ambiguous statements. Second, physical injuries such as burns, scars, or limitations can easily be demonstrated. Third, a videotape may have a greater effect on a jury if portions of the deposition are introduced at trial as evidence. Finally, a videotape can serve as a more effective substitute for a party who cannot testify at trial, such as an expert witness from another state or a witness who is too ill to be brought to the courtroom. If a witness dies unexpectedly before trial, a videotaped deposition can be admitted in lieu of live testimony because the deposition was taken under oath and the opposing attorney had the opportunity to cross-examine the witness.

Another advance in technology is the ability to take depositions by telephone. Telephonic depositions are allowed under the federal rules and are acceptable in most states. The procedures for a telephonic deposition are the same as for a regular deposition, although it is preferable (and sometimes required) that the examining attorney state for the record that the deposition is being taken over the telephone. A telephonic deposition can occur with the attorneys and the deponent in three different sites; in any case, federal and state rules stipulate that the judicial district within which the deponent is located is the official site of the deposition.

Another technology used for depositions is videoconferencing, where sound transmitters and receivers are combined with video cameras and monitors, allowing the attorneys and deponents to see each other as a deposition proceeds. Videoconferencing makes the examination of exhibits easier and also helps reduce confusion among the participants that may result from ambiguous or unclear verbal responses.

5 Reports

5.1 INTRODUCTION

The presentation of convincing testimony is the ultimate goal of any witness. This can be accomplished by understanding the environmental forensic process. The purpose of this chapter is to give an introduction to the report writing process.

In the final analysis, most settlement conferences and mediations boil down to negotiating dollars, and technical issues usually represent a factual foundation, posturing, and window dressing (unless there really was only one side to the technical issues). Even though the majority of environmental cases settle, it is important to anticipate and therefore prepare for trial testimony early. Failure to do so can lead to disaster if one is suddenly confronted with unexpected demands, schedule changes (which almost always occur), or a settlement impasse.

Generally, reports (usually written reports) form the basis for pretrial preparation, settlement negotiations, and testimony during a trial. The submittal of requested reports may lead to a decision not to call the expert witness or to effect a settlement and hence prevent a trial. If a written report is requested, the expert should incorporate only the necessary information used to reach conclusions and opinions.

Many state-court trial attorneys prefer no written reports from experts; they want to avoid pinning the expert down until all the evidence is collected. However, federal courts now require written reports. In fact, there are several examples of problems or issues that can develop with written reports—for example, premature conclusions, inconsistencies, and scope of analysis incongruities. However, in some situations reports are particularly helpful. These include (1) reports for settlement; (2) reports for mediation, arbitration, and minitrial; and (3) declaration reports for summary judgment motions.

For the scientist or engineer appearing as an expert witness, it is best to assume that a report will be required. There should be no doubt, because this should have been one of the topics discussed during the early stages of retention, after the interview. Furthermore, under the Federal Rules of Evidence, opposing counsel (i.e., the cross-examiner) may require scientific and engineering experts to disclose in detail the underlying facts and data for their opinions. Under these circumstances, a party offering testimony more related to junk science than to real science risks a very embarrassing cross-examination.

A judge can instruct the jury that the testimony of an expert is to be evaluated like that of any other witness, and the jury is free to reject the testimony entirely if it is not credible. Many maintain a healthy respect (sometimes a healthy skepticism) of expert testimony and comply with the intent of the judge's instructions. This may seem unfair to many scientists and engineers (and attorneys), but it is an effective means to ensure the integrity of the expert; with cross-examination and presentation of related evidence, it will support the admission of expert testimony.

However, this requires effective preparation by the expert, and the 1993 amendments to the Federal Rules of Civil Procedure, which mandate the disclosure of detailed expert reports, should facilitate preparation for cross-examination. Moreover, in federal court, a further protection exists in the practice of allowing the judge to question witnesses called by the parties in order to assist the judge's assessment of confusing or misleading evidence. Therefore, the scientist or engineer who appears before the court as an expert is required to submit a written report prepared in accordance with the guidelines established by the court or by the relevant federal authority.

In a trial in which an expert witness participates, not only the testimony of the expert witness but also reports are critical to success. Reports, whether oral or written, can range from minimal to extensive. Many attorneys request reports that are not opinion reports. This aspect depends on the needs of the attorney and must be discussed very early in the arrangement. By no means should the expert witness ever send anything in writing until a request is made by the attorney.

In fact, before proceeding it is worth remembering that testimony is a sworn oral report by a witness of information relating to a legal dispute. However, the oral reports described in this chapter relate to those specifically prepared and presented by an expert witness to the judge and jury or to the mediator or arbitration panel.

The subject of an oral report can be broached at any time during the attorney–expert contacts, but it is best laid on the table as soon as possible. At the time of assignment of the trial, the expert may be asked (if he or she has not already been asked) to review the facts in the case as presented at that time and prepare a preliminary report. This report may be an oral report to give the attorney enough information to make a decision on the question of settlement or trial. The expert's analysis of the facts and opinion may be the basis for a settlement conference and easy disposition of the case.

When it appears that settlement is not going to result from these efforts, the expert will more than likely be asked to proceed with a thorough and complete investigation that ends with a written report giving an opinion as well as the facts forming the basis of that opinion. Rendering a written report is the most significant time restraint that will have to be met in completing the investigation at the time specified by the court in its rules or in its orders.

The expert witness must always keep in mind that all of his or her reports and written communications may be shown to the opposing counsel. For this reason, it is best for communication between the expert and attorney to occur only in person or by telephone. Written material (including any faxes) sent to the attorney may not be designated by the court as confidential and is therefore discoverable. Opposing counsel has the right to examine all documents, exhibits, or material relevant to the case at the time of discovery. Some courts have ruled that even the notes made by the expert witness that are relevant to the case may be given to the opposing counsel during the discovery process.

However, from the beginning, even at the time of the first contact with the attorney, the scientist or engineer should anticipate that an oral report may have to be presented to the court. This report is not necessarily oral testimony. Testimony is information presented to the court under direct examination from the attorney. An oral report requires that the expert stand before the court and present his or her findings orally in a logical and convincing manner. A written report may form the basis for the oral report, but making an oral report is not merely a matter of reading the written report before the court. The judge will dictate the format of the presentation, and the expert is advised to be ready for such an assignment. On the other hand, the format and the means of oral presentation may be covered in the rules of procedure of the court.

Finally, if an expert witness prepares a report and believes (or knows) that it is incomplete or inaccurate without some qualification, that qualification should be stated in the report. On the other hand, if an expert witness considers that his or her opinion is not a concluded opinion because of insufficient research or insufficient data or for any other reason, this must be stated when the opinion is expressed. In addition, an expert witness who, after communicating an opinion to the attorney (i.e., the party who retained him or her), changes his or her opinion on a material matter should provide the attorney (the engaging party) with a supplementary report to that effect, which must contain such additional information.

5.2 EXPLAINING SCIENCE AND ENGINEERING TO JUDGES AND JURORS

Laypersons in the courtroom (i.e., the judge and jurors) will hear many conflicting claims during the course of a trial, and the expert should counteract such potentially confusing messages by writing the report in plain and clear language. Above all, the expert should remember that the views expressed in the report are his or her own and do not necessarily reflect those of the law firm or organization that has retained the expert. The report (oral or written, and the former may be in the form of a presentation or oral testimony) is the means by which the scientist or engineer is allowed to explain technical issues to the judge and jury.

However, recent changes in the law of evidence have given scientifically untrained judges the power to determine the reliability of scientific interpretation of scientific observations. Effective interdisciplinary communication requires now, more than ever, that scientists and engineers recognize that scientists and engineers, medicine, law, and politics use different tools and methodologies and pursue different goals; that scientists and engineers become better aware of the strengths and limits of their own and the legal profession; and that legal decision-makers meet a higher standard of scientific education so that the latter understand that chemical laws have universal validity and are not merely a consensus of a community of specialists.

The *Daubert* decision requires trial judges to determine the relevance of evidence and the reliability of the data on which the expert relies. Trial judges must also determine whether the methodology used by the expert and the opinion expressed by the expert are reliable or whether there is too much of an analytical gap in the expert's reasoning and too much extrapolation. This poses a challenge for the judge because expert testimony is only allowed when a lay person's knowledge and reasoning are insufficient to understand a phenomenon or issue and the judge, who normally decides a case by determining what is best for society, is poorly equipped to interpret scientific data because natural laws are not governed by the principle of justice.

On the other hand, the task of the jurors is to adjudicate disputes on the basis of the laws and standards of the local community and jurisdiction. Jurors reflect the diversity of the local society.

The problem of explaining science and engineering to jurors is the same as explaining science and engineering to other nonscientific and nonengineering audiences. Jurors do not usually adjudicate scientific facts; they decide what testimony is more relevant, plausible, and credible. Furthermore, the interdisciplinary coordination of experts is difficult because each of these professions uses different tools and methodologies and pursues different goals.

Scientists and engineers have a unique place within the education system in that they educate students not only to plan experiments carefully, identify and isolate parameters, and observe but also to avoid shortcuts based on intuition and superficial perceptions. In fact, science and engineering occupy a unique position in modern society in that they are embraced as a point of reference in many arenas of activity that are not of a scientific or engineering leaning. Moreover, the impressive success of science and engineering in establishing their authority has depended upon the distinct commitment of scientists and engineers to the goals and procedures of science and engineering.

Thus, scientists and engineers are called to contribute to decisional areas, including the judicial system, which in turn must struggle to incorporate science and engineering into its own distinct systems of thinking and acting. If scientists and engineers want scientific laws to be respected in the legal forum, they need to explain the applicable laws in terms that nonscientists and nonengineers can understand.

Interdisciplinary collaboration offers a small window into difficulties that arise when usually bounded activities are drawn together. Even among persons with much shared training—scientists and engineers—friction is common among subspecialists. Tension arises as each gives priority to his or her team's assessments, methods, and perceived scope of competence. This divergence is relatively small when compared to activities where purposes are fundamentally different, such as when we contrast science and engineering with other professions.

Junk science and engineering are only a threat when science and engineering are not credibly and effectively presented. The credibility of a scientist or engineer depends on candor, mastery of the subject matter, preparation, ability to communicate, and demeanor of the expert and the party's attorney. The jurors can sense whether or not a litigator understands the science and engineering; a good scientist or engineer can help win a case if he or she understands the strength and shortcomings of a case and candidly communicates both to the jurors.

A frequent argument among scientists and engineers, especially researchers, is that science and engineering cannot be adequately explained to judges and jurors because they are lay people who lack the preparation to evaluate science and engineering. However, juries and judges do not determine scientific facts; juries determine which evidence is more relevant to the case and more credible.

It is important for researchers to recognize that the overselling of science and engineering creates havoc for those who have to live with the consequences and that not only lay people and lawyers but also engineers and clinical physicians believe that scientific literature is not always more trustworthy than commercial product advertisement.

Explaining science and engineering to judges and jurors requires thorough, specialized knowledge; good oral and written communication skills; an appreciation of the cultural differences that separate scientists and engineers from lawyers and jurors; and at least a superficial understanding of the rules of communication that govern legal proceedings.

5.3 THE ORAL REPORT

An *oral report* is given in the form of testimony in a deposition or on the witness stand before a court of proper jurisdiction or other trier of fact. In other words, an oral report is a word-of-mouth report that is made by the expert to the judge and the jurors. In addition, a preliminary oral report is often the first presentation made by the scientist or engineer to the attorney. Such a presentation must be focused, understandable, and professional. However, this is not the major point of this section; rather, it concerns the presentation of an oral report to legal proceedings (mediation, arbitration, or trial).

The judge may request by-attendance-in-court oral reports from each of the experts involved in a case. Alternatively, although less likely, the judge may request oral reports from the experts by telephone, allowing each expert to speak on different occasions and blind to the other expert reports. Many judges maintain contact with the attorney and opposing counsel and may request oral reports from them on a semiregular basis.

However for the expert witness, oral reports should not occur only once during a project but, rather, regularly insofar as the expert needs to keep the attorney up to date. Such reports can be made by telephone or in person. As part of the report to the attorney, the expert informs the attorney what has been done and the conclusions. If the expert has formed an opinion, the attorney should be advised of this opinion. He or she may or may not be looking for certain things from that particular aspect of the evidence, so it is important to work with the attorney in doing that.

Appearing in court can be a daunting experience for any scientist or engineer. In fact, the scientist or engineer usually requires training and experience specific to the development of expert witness skills and expert witness testimony, to assist in litigation with an oral report. Preparation usually includes training or experience in the areas of (1) trial preparation, (2) direct testimony, (3) cross-examination, and (4) use of demonstrative exhibits and learned treatises.

Most scientific and engineering disciplines require that the respective scientists and engineers make presentations before their peers as part of intracompany meetings or part of national association meetings for career development. Such audiences can be very impolite in the manner in which the speaker is questioned. Thrusting questions and running commentaries on the work performed and reported by the scientist or engineer are the order of the day. At such meetings, chivalry (as it is supposed to exist) is quite dead!

However, meetings of this type where the scientist's or engineer's peers have been decidedly unfriendly, to say the least, form the basic groundwork for an oral presentation to a mediator, an arbitrator, or a judge and jury, but such meetings are not the total answer. This can only be achieved by working with the attorney to build upon the inherent and acquired talents of the scientist or engineer.

The oral report may be much more effective than the written report. The judge and jury can read the body language of the presenter and have the ability to determine the sincerity of the expert. Proceedings (mediation hearing, arbitration, and trials) can be won and lost based on such reports. In addition, the expert should never forget that he or she can be questioned by hostile opponents based on the content of an oral report.

The courtroom is the attorney's domain. The expert is well advised to listen to him or her and take heed of the potential benefits and pitfalls of oral presentations.

5.3.1 The Nature of the Report

As already stated, for the current context, an oral report includes any information in relation to a dispute or lawsuit given verbally by an expert to a judge. Oral reports or presentations are sometimes used in addition to the formal written report or opinion letter. The oral reports are useful to keep the client informed of the progress of the investigation and to summarize any preliminary findings. In fact, the purpose of the oral report is to provide rapid additional information to the court to enable the judge to make a decision at an urgent hearing. The oral report will advise the judge of the findings to date; in the case of bench trial, he or she might decide against proceeding further because of the lack of any additional benefits. Oral reports may include (1) verbal information; (2) copies of recent reports, memos, and relevant letters; and (3) any information specifically requested by the judge.

In summary, an oral report should provide factual information in an understandable manner (Table 5.1) to assist the trier of fact to assess the merit of one side or the other in the dispute or lawsuit and to determine a course of action. Plaintiffs can benefit because meritorious cases can be shown to be on sound technical ground, and this can provide the basis for a realistic opinion on settlement issues. On the other hand, defendants can benefit if the oral report shows a backbone for a solid defense as well as a realistic opinion on settlement issues.

TABLE 5.1
Various Aspects of an Oral Presentation

Aspect	Comment
Introduction	Have a complete and well-organized overview
Completeness	Address all of the required elements
Organization	Be well organized Move from general topics to specific details without confusion
Understanding	Allow everyone to understand the presentation
Oral skills	Be easily understood Maintain eye contact with the judge and jury
Graphics	Emphasize the main points Be understandable and professional
Technology	Reduce technology to a nontechnical, understandable level Do not overwhelm the audience
Questions	Answer questions confidently, clearly, and accurately

5.3.2 Presentation of the Report

One of the most neglected fields by expert witnesses is that of presentation techniques. Bad presentations are also one of the most damaging, discouraging, and dispiriting aspects of any case. Presentations may be so bad that a sure-fire-winner case can turn out to be a loser because of the bad presentation made by one of the experts.

For the presentation, the expert should be aware of the situation. The classroom situation can be intimidating and dramatic. The expert should be prepared to respond adequately, in which case reading from a script by the expert (no matter how well the script is prepared) is not recommended. The expert should believe in what he or she is doing. A visibly detached attitude—reading a script as if it did not belong to or had not been prepared by the expert—may well be a source of irritation for the judge and the jury, if the jury is present.

In the United States, the language of the court is English. If English is not the mother (or father) tongue of the expert, a typewritten, double-spaced, twelve-point font size copy of what the expert is going to say should be prepared. This does not mean that the expert should read everything from the paper but, rather, that he or she should have it there and ready in case of necessity.

The expert should determine beforehand (from the attorney) the amount of time allotted for the presentation. The judge may give a time limit of, say, twelve to fifteen minutes for the expert to include all of the salient facts. The expert should adhere to the time limit. As a general rule, one letter-size page, double-spaced in twelve-point font, is equivalent to approximately three minutes' speaking time. The expert should read the text aloud at home and note the time taken. The speaking style should be slow and clear and allowance made for pauses.

It is advisable to use and prepare media other than the expert's own voice; this should be integrated into the presentation. In this sense, the expert should make certain that the necessary technical equipment is in the courtroom (or ready for the judge's chambers) in time and in working order. If the expert is to use a blackboard, he or she should make sure it is clean and, if possible, have some of what is going to be used on the blackboard already written out before the start of the presentation. If the presentation is to take place in the judge's chambers, an oral presentation only, without equipment, may be the easiest to do.

The most important part of any presentation is the expert's voice. Everything else should be an auxiliary to the expert's role and function in the presentation. A good presentation should allow for a discussion by or questions from the judge. The presentation is for the judge, and he or she should be treated with the courtesies that one would give to the chief executive officer of any large major company.

As far as mimicking, gesturing, positioning, and movement in the courtroom are concerned, there is no one form that fits all. Some experts will feel (and look and come across) better while sitting at a desk when speaking (being seated is at the discretion of the judge). Some experts use gestures very sparingly, and others will gesticulate agitatedly while pacing to and fro across the room, enacting their presentation to the hilt and further and still not looking ridiculous.

Ultimately, the expert's performance during the presentation is an issue of personal taste, but the expert should make sure that it is his or her own choice of style. The positioning and style of presentation must be a conscious decision, based on the expert's own abilities and on the demands of the situation. Most of all, it must be in the expert's own style so that the presentation will come across convincingly.

It is in order to offer the judge, through the attorney, a handout. Of course, there should be an evident relation between what the expert is saying and what is on the handout. Ideally, the handout (1) summarizes and highlights the main points of the presentation; (2) presents, in written form, citations of the sources; (3) may provide additional information that is not part of the presentation; and (4) should be in the hands of the judge *before* the expert starts speaking.

5.3.3 Use of Graphics

The use of graphics or visual displays, subject to the permission of the judge, can enhance an oral report and the understanding of the technical issues by the judge and jury. The scientific or engineering evidence provided by the expert may be hard to follow; thus, an oral report can be used to portray the concepts particularly effectively, especially when supplemented by professionally drawn charts, graphs, drawings, or models.

The expert provides an effective presentation by allowing the judge and jury to view the oral presentation in a different way. Graphics allow the judge and jury to observe the important points in the correct context but as stand-alone items. Graphics or visual displays mean nothing and are not usually considered to be evidence. They merely serve as testimonial aid and are not original evidence of what occurred. However, the main issue to be decided before the use of such materials is whether or not the graphics (drawings, maps, charts, models) will aid the judge and jurors in understanding the expert evidence as presented in the oral report.

If the expert decides to use graphics as an integral part of the oral report, he or she should at least consider, if not adhere to, the following rules of thumb: (1) the graphics must represent the underlying data, (2) the graphics are used because the amount of underlying data is sufficiently voluminous to make examining the data inconvenient for the judge and jurors, (3) the underlying data must be admissible as evidence, and (4) the underlying data must have been disclosed to the other side. This last item is an issue for the attorney to handle.

Furthermore, any graphics that are used by the expert will, most likely, be marked as trial exhibits. For the duration of the trial, the package will become the property of the court. Because the package is then subject to close inspection, any errors or contentious issues contained within the graphics package will be eagerly pounced upon and devoured by opposing counsel.

As a reminder for the expert, all graphics should be seen and examined carefully by the attorney for relevance, correctness, and any other reason the attorney can call to mind. Furthermore, it is very worthwhile for the expert to make a trial run (no pun intended) of his or her oral presentation in front of the attorney and anyone else he or she considers necessary to be included in the audience. This gives the expert the chance to rehearse his or her words and the attorney the chance to check the intimate details of the presentation for errors of fact—major or minor, they should not be present. Failure of the attorney to do so can result in a lost case. As a reminder, the attorney or opposing counsel should not let his or her respective experts loose with uninspected graphics for the same reason.

5.4 THE WRITTEN REPORT

If a report is requested, the expert witness should be sure to get guidelines from the attorney. Furthermore, the report should list the facts as the expert witness knows them, and any information obtained from other sources should be referenced. If the report is related to the opinion, above all, the opinion expressed must be clearly stated to be the expert witness's own opinion. The attorney may try to influence the expert witness in some phase of the report to enhance the attorney's position and court strategy. This must be resisted at all costs.

An expert's report is one of the most important services provided to retaining counsel. A well-written report is immensely helpful to retaining counsel and may well lead to future referrals and the ability to charge premium fees. A poorly written report can and will be used to impeach the expert in the case at hand and future cases for years to come.

The specific content of any expert report is determined by the type of dispute or lawsuit, but, whatever the issue, the ultimate purpose of the report is to help the judge and jurors make a decision. Briefly, the report should contain comprehensive and complete statements with substantive evidence to support all opinions or conclusions expressed. Expert reports should not be written to impress other experts; they are written to inform the judge and the jury. Thus, the excessive use of obscure or unexplained technical terms does not inform the reader. Just stating opinions and conclusions is not adequate. The most valuable information is that used to substantiate the opinions or conclusions. Without substantiation, opinions and conclusions are just statements; with substantiation, they become of value to the client.

The data and other information relied on as the basis for opinions and conclusions must be identified. Credibility comes from using appropriate sources of information. These sources, whenever possible, should have firsthand information. If secondary sources are used, they should be identified as such. Copyright restrictions must be considered when using information published by other than the report author. Exhibits used to support opinions and conclusions must be succinctly identified and appended to the report. How the exhibits are used and reported is determined by the writer based on overall clarity and their contribution to understanding the report.

The authenticity and validity of the report must be acknowledged by the signature of the writer. Facsimile signatures and signature stamps should not be used. An otherwise professional report with something other than an original author's signature dramatically reduces the professionalism of the report.

5.4.1 Purpose

In the practical world of the courtroom, an expert report conveys information, conclusions (hence, opinions), and recommendations from an expert who has investigated the cause of a dispute or litigation in detail. The information is needed for the specific purpose of helping the judge and jurors reach a decision. A report of this kind differs from an essay in that the report is designed to provide information that will be acted on rather than to be read by people interested in the ideas for their own sake. Because of this, the expert report has a different structure and layout.

Written expert reports often ensure greater accuracy than oral reports and can serve as a basis for drafting individualized interview questions. However, if this part of the process is confidential, written reports may also increase the chance that investigative information will find its way into the public arena. When a judge feels strongly that confidentiality cannot be maintained once reports are reduced to writing, he or she may want to rely on an oral report.

Because an expert report conveys information, producing one should logically be organized around the four following stages: (1) framing the issues and planning, (2) information gathering—researching the project, (3) analyzing the information, and (4) writing the report.

The sources available to and used by the expert will be determined by the aims and scope of the report. Sources, for the purposes of the current context, are materials and information gathered during discovery, as well any relevant data, information in books, technical journals, and other reports. The expert should ensure that the information used is relevant and the source must always be referenced. For a written investigative report, the report should indicate the reliability of the information, and the expert must be aware of the scrutiny a written report will undergo.

Written reports form the basis for pretrial preparation, settlement negotiations, and testimony during trial. They may lead to a decision not to call the expert witness or to reach a settlement and hence prevent a trial. Reports may be a few paragraphs or voluminous. If a written report is requested, the expert should incorporate only what is necessary. Gratuitous and unimportant comments are to be avoided. If a report requires permission, consent, or a waiver of confidentiality, the expert should insist that counsel get the proper authorization. The expert must be aware of the due date for the report and to whom it is directed.

5.4.2 Writing the Report

It is not sensible and may even be frustrating and stressful for the expert to leave all of the writing until the end of his or her research. There is always the possibility that it will take much longer than anticipated and there will be insufficient time, thus leading to an all-night session (or several all-night sessions) during which clarity of thinking has departed for the day.

It is wise for the expert to begin writing up some aspects of his or her research as the work proceeds. At this stage, the report does not have to be written in the order that it will be read. Any order of writing that suits the expert is acceptable. The use of a word processor makes it very straightforward to modify and rearrange what has been written as the research progresses and ideas change. Some experts print the early drafts as hard copy with one paragraph to a page; using this technique to shuffle the paragraphs page by page is relatively easy. The bottom line is that the expert should choose the method most comfortable and familiar to him or her. Within the writing timetable, the expert sets deadlines for different sections of the report while keeping in mind the deadline set by the attorney for delivery.

5.4.3 Format

The expert should follow the formatting guidelines provided by the attorney at the time the request to write the report was made. Usually, these guidelines have been passed down from the court or the attorney has found them acceptable to the court when reports were written by experts on prior cases. To prevent misunderstandings, the expert should request a list of questions or an outline of the issues the report should address from counsel. As a minimum, the report should identify the reason for the report, the matter to be explored, the witness's observations and rationale, and other significant information and sources.

An expert's written report must give details of the expert's qualifications and of the literature or other material used in making the report. All assumptions of fact made by the expert should be clearly and fully stated. The report should identify and state the qualifications of each person who carried out any tests or

TABLE 5.2
Example of a Cover Page of an Expert Report

REPORT OF INVESTIGATIONS INTO
THE BEHAVIOR OF XYZ PRODUCT
DURING CONDITIONS OF SERVICE

Submitted by

Dr. James G. Speight

CIVIL ACTION NO.

ABC Inc.,

Plaintiff,

vs.

XYZ Inc.,

Defendant

IN THE UNITED STATES DISTRICT COURT

DISTRICT OF LMNOP:

Date

experiments upon which the expert relied in compiling the report. When several opinions are provided in the report, the expert should summarize them. The expert should give the reasons for each opinion.

The cover page or title sheet (Table 5.2) should succinctly identify the nature of the report. Within the report there should be a comprehensive table of contents (Table 5.3), which allows the reader to identify the issues covered by the report immediately. Depending on the preference of the client and the length and complexity of the report, the various sections can be expanded or shortened as necessary. The client may prefer to base the format on that of the American Society for Testing and Materials (ASTM E620: Standard Practice for Reporting Opinions of Technical Experts). Whichever format is used, the report needs to deal with the issues in a clear, concise, and convincing manner (Table 5.4).

This practice standardizes elements of the expert's report, which will make the report understandable to the intended recipient and focus on the technical aspects germane to the purpose for which the opinion is rendered. This standard practice covers the scope of information to be contained in formal written technical reports that express the opinions of the scientific or technical expert with respect to the study of items that are or may reasonably be expected to be the subject of criminal or civil litigation. If hazardous materials, operations, and equipment are involved, the standard does not purport to address all of the safety concerns associated with their use. It is the responsibility of the expert who uses this standard to consult and establish appropriate safety and health practices and determine the applicability of regulatory limitations prior to use.

TABLE 5.3
Table of Contents of an Expert Report

TABLE OF CONTENTS

5.4.4 Facts, Data, Opinions, and Conclusions

The report should include all *facts* and *data* relevant to the opinion rendered. The expert should include whatever detail he or she considers necessary. When data from examination or testing (of a product) are included, the sources of the data must be cited and the person providing the data must be consulted to determine his or her opinions. In addition, each piece of equipment used in the examination or analysis should be identified, and there should also be acknowledgment of the means by which any instruments are calibrated—especially if this is a standard procedure before instrument use. Such information may be included in an appendix to the report. The report should identify each *opinion* and *conclusion* as well as the expert's reasoning for each opinion and conclusion.

TABLE 5.4
Guidelines for a Written Report

(a) Content

Content Areas	Comment
Problem	Must be clearly described
Content	Includes definitive statements and conclusions
Illustrations	Include illustrations where necessary Detailed drawing if required Correspond to information provided in the text
Adaptations	Explain accommodations relating to the five human senses Explain how these senses are adaptations to the environment
References	Make appropriate use of references to give credit due to others Include four or more references

(b) Style

Style Areas	Comments
Clarity	Easy to read Clear and concise Provide examples as appropriate
Accuracy	No errors of fact
Completeness	Statements that are relevant to the case Explain key ideas fully
Depth	Address main factors that make the topic important Show evidence of using several major relevant resources
Punctuation, grammar, spelling, and appearance	No punctuation errors Few grammatical errors No spelling errors Professional and well produced, using appropriate font, font size, line spacing, and border areas

5.4.5 Other Items for the Report

If the final report is the work of one individual, his or her *signature* should be affixed at the end of the section listing the conclusions. If the report is a joint effort, the signature of each expert should be included. However, if separate opinions relative to different issues are involved, it is necessary to identify exactly the opinions and conclusions of each expert. If submittal of a joint report is mandated to emphasize opinions and conclusions of a team of experts, it is preferable that each expert contribute his or her own individual section of the report.

On the other hand, submittal of a joint report in which the sections written by the various experts are not well defined can be fraught with risk. A joint report gives opposing counsel the opportunity to confuse the joint authors and confuse the judge and jurors. Rather than risk such an outcome, it is better to consider the preparation of reports written by individual experts.

Each expert participating in the development of a report should give a detailed outline of his or her *qualifications*. The qualifications should include education and training that bear upon the expert's ability to render opinions relative to the issues under consideration by the court, identification of the relevant educational institutions and their location (city, state or province, and country), dates of attendance, and degrees or honors received.

Experience that is directly relevant to the expert's ability to render service is also necessary. This should include a list of relevant projects on which he or she has been engaged (firm, city, state or province, and country), the number of persons in his or her responsible charge, dates of employment, and titles held.

Professional affiliations, including organizations in which the expert holds membership; significant service to these organizations and dates involved; and any honors bestowed or awards won are also of value. Other *professional activities*, such as a list of papers published, books written, and lectures delivered, are also to be included. Elimination of any professional activities on a selective basis other than on a calendar basis may give opposing counsel cause to question the expert's qualifications and credibility.

Appendices can be used for copies of documents associated with the report or investigations, tables, charts, photos, drawings, test results and data, as well as other documents pertinent to the expert's opinions and conclusions.

In summary, included in or attached to the report should be (1) a statement of the questions or issues that the expert was asked to address, (2) the factual premises upon which the report proceeds, and (3) the documents and other materials that the expert has been instructed to consider. Usually, such items are included at the front of the report under a section designated "Assignment."

If, after exchange of reports or at any other stage, an expert witness changes an opinion after having read another expert's report or for any other reason, the change should be communicated in a timely manner (through legal representatives) to each party to whom the expert witness's report has been provided and, when appropriate, to the court.

If an expert's opinion is not fully researched because the expert considers that insufficient data are available, or for any other reason, this must be stated with an indication that the opinion is no more than a provisional one. When an expert witness who has prepared a report believes that it may be incomplete or inaccurate without some qualification, that qualification must be stated in the report. In addition, the expert should make it clear when a particular question or issue falls outside the relevant field of expertise. However, this latter point should have been addressed by the expert and the attorney at one of the early contact meetings.

Finally, when an expert's report refers to photographs, plans, calculations, analyses, measurements, survey reports, or other extrinsic matter, these must be provided to the opposite party at the same time as the exchange of reports. The report should state any conclusions the witness may have reached based on the observations and information. The expert should take a copy of the report with him or her when attending court to testify.

The expert will, in all probability, be questioned by the opposing counsel as to whether the report has been in any way influenced by the attorney or someone else who wants a favorable outcome for the case. The expert should always remember that it is not his or her case; the expert is not the attorney.

In addition to reports, the handling of personal notes must be discussed early in the process. All experts should review their habits with regard to notes: do they keep and file notes? Do they take notes to refresh their memories of conversations and of material read in books? Do they take notes, review them, and then throw them out? All of these questions must be discussed early on. Experts need guidelines for every case because different attorneys have different ideas about how to deal with these topics. Each time the expert writes anything, that section should be dated.

The expert must think ahead because any notes made (for any reason) will invariably be given to the opposing counsel during discovery or if subpoenaed. Therefore, any notes should also be signed at the bottom of each page and dated. After the report is written, the notes can be destroyed.

The expert must be prepared for the opposing counsel to ask him or her to furnish all material on which the expert has relied to come to a conclusion. This material may include photocopies of references and even copies of the pages of books used. Photocopies of the title pages and the pertinent sections are sufficient. The expert should not submit entire books because they may not be returned. If the actual books are provided, the expert should make it clear to both the attorney and the opposing counsel that they must be returned to the expert after the trial, but this does not always happen. It is wise to make notes and records of everything provided to the opposing counsel. Some trials take more than a year to get on the judge's calendar; it is easy to forget over that time exactly which books and returnable materials were given to the opposing counsel.

5.4.6 Understandability

The final written report must be professional. Handwritten reports must be avoided; use of a good word processor is in order. The report should have a cover sheet giving the following information: the expert's name, address, phone number, fax number, and e-mail address; the names and addresses of the client and the attorney; and the title and court number of the case.

The first page of the narrative should review the problem and the claim of injury. These statements should be followed by the expert's evaluation and a clear statement of the opinion or conclusion derived by the expert. If necessary, pertinent references should be cited; if applicable, charts, diagrams, and photos (or photocopies) should be appended.

Specific references to recognized journals are the most acceptable. Opinions of other experts can be used if they are explicitly recorded in a deposition or scientific publication; on the other hand, opinions based solely on the conclusions of other scientists are not acceptable. It is difficult to determine how circumstantial evidence can be used; the guidance of the attorney is important here. The expert must resist the request of an attorney for the expert to introduce inadmissible evidence under the pretense of explaining the reason for the opinion.

Above all, the opinion stated by the expert must be based upon known facts and scientific probability. In light of the Supreme Court's *Daubert* decision, the expert must demonstrate the validity of the analysis and conclusions. Some states require that the written opinion be given to the jury.

Finally, the expert must be prepared to have what he or she says and writes contradicted by other experts, including friends and colleagues. It is nothing personal but, rather, just part of the job. The expert should make sure that his or her report contains information that is generally accepted in his or her field, and the information should be supported with references. Even then, there are genuine differences of opinion in any discipline, and the expert must expect to be contradicted. This is all part of being an expert witness.

In the final analysis, the expert's opinion is not a fact. Once the expert has stated an opinion, he or she should stand by it. If badgered, the expert should hold his or her ground and say, "That, sir [or ma'am], is my expert opinion."

5.5 CHARTS, FIGURES, AND VISUAL AIDS

The use of visual aids in court is very dramatic if done right. They must be accurate. A good visual aid can be very effective in getting the expert's points and ideas across to the jury. At all times, any visual aid should be covered until used and then left uncovered. Charts, graphs, diagrams, maps, and photographs are often used. If appropriate, models can be made and exhibited; some experts use videotapes. These can be rather boring if not made well; if done professionally, they can be a powerful teaching tool. One advantage to videotapes is that they can be replayed. Virtual reality will, no doubt, one day become a tool for the expert to use in the courtroom. However, an expert should not make the material appear to be too expensive.

Trying to take the jury's mind away from the substance so that they only see the picture could backfire. For example, if vivid computer animation is used so that the jury could follow the path of an accident. The jury may not be impressed and not rely on the recording.

If the expert makes some of the visual aids himself or herself, they will carry much more impact with the jury than, for example, photographs made by a professional photographer. However, any photographs used must be of very high quality and enlarged enough to be seen in the courtroom. Visual aids must be very simple, clear, and not crowded. One or two ideas (or at the most three) should be presented per chart. When using charts or diagrams, the expert must be careful to stand on the side and not block the view of the jury. If the diagrams or charts are to be marked, the expert should use colored markers and make all notations big. Charts can be used again for closing arguments.

The conclusion should be given orally; only then is the final sheet placed on top. Needless to say, the overlays should not obscure the information given on the sheets below. All visual aids must be created so as to aid the jury in understanding the science of the case. The decisions on how to make and present the visual aids should take into account the fact that these exhibits will go to the jury deliberation room without the expert.

All this advice on visual aids is predicated on the assumption that the attorney has introduced the material into evidence and the court has assigned a number to it. If this is not the case, the judge may not permit the expert to use the material. When referring to any visual aid, the expert should use the number assigned to it when it was placed in evidence.

If at all possible, the expert witness should avoid doing a laboratory experiment in the courtroom and should never bring a rodent into the courtroom. If necessary, an experiment can be done in a clean, orderly laboratory and videotaped; this precaution will prevent an experiment from failing at the wrong time in front of a jury.

6 The Predeposition and Deposition

6.1 INTRODUCTION

As noted in Chapter 1, a controversy before a court or a lawsuit is commonly referred to as *litigation*. If the dispute or controversy is not settled by agreement between the parties, it will eventually be heard and decided by a judge or jury in a court. Litigation is one way that people and companies resolve disputes arising out of an infinite variety of factual circumstances. Scientific and engineering expert witnesses are used more often in civil litigation; therefore, the focus of this text is, for the most part, the use of such experts during civil cases.

At some point, a person or company may be either a defendant in a lawsuit or contemplating filing a lawsuit against someone else. A lawsuit typically (but not always) proceeds through four basic phases:

1. the filing of initial pleadings;
2. the discovery phase;
3. the deposition phase; and
4. the post-trial or appellate phase.

This chapter deals with the first three of these phases and gives an indication of the procedures involved in each. In fact, the procedures or events leading up to and including the trial fall under the various rules of procedure or procedural law.

Procedural law is law that governs the machinery of the courts and the methods by which both the state and the individual (the latter including groups, whether incorporated or not) enforce their rights in the several courts. Procedural law prescribes the means of enforcing rights or providing redress of wrongs and comprises rules about jurisdiction, pleading and practice, evidence, appeal, execution of judgments, representation of counsel, costs, and other matters. Procedural law is commonly contrasted with substantive law, which constitutes the great body of law and defines and regulates legal rights and duties. As an example, *substantive law* is the means by which two parties enter into a contract, and *procedural law* is the means by which one party would allege a breach of that contract and seek the help of the courts in enforcing the agreement or collecting damages because of breach of the agreement.

Thus, procedural law is a means for enforcing substantive rules. There are different kinds of procedural laws, corresponding to the various kinds of substantive laws. Criminal law is the branch of substantive law dealing with punishment for offenses against the public and has as its corollary criminal procedure, which indicates how the sanctions of criminal law must be applied. Because the object of judicial proceedings is to arrive at the truth by using the best available evidence, there must be procedural laws of evidence to govern the presentation of witnesses, documentation, and physical proof.

Pretrial matters such as pleadings, motions, and discovery that involve the presentation of witnesses, documentation, and physical proof are governed by various complex *rules of procedure*. This stage of the dispute can dictate strategy and how the litigation will progress.

6.2 THE PLEADINGS STAGE

Civil litigation is initiated when one disputant (the *plaintiff*) files a *complaint* (also known as a *statement of claim, petition,* or *declaration*) with the court having jurisdiction. In the complaint, the complainant identifies allegations made against the *defendant* and the nature of relief sought, usually in the form of monetary damages. The trial is what most people think of when they hear the terms *lawsuit* or *litigation*; however, most of the work is done during the pretrial phase, which includes preparing and filing pleadings and motions and exchanging discovery results.

Pleadings are documents that outline the parties' claims and defenses. In a motion, a party requests that the court take a specific action. Motions can cover a wide variety of issues, from asking the court to compel a witness's testimony to requesting that the court enter a protective order so that sensitive information is kept confidential.

A lawsuit starts with the *summons,* which gives notice of the suit to the person or entity being sued, and the complaint. The summons is usually delivered in person by a marshal, sheriff, constable, or some other process server. The summons identifies the defendant and action being taken against him or her, the name of the court, the name of the plaintiff, and the name and address of the plaintiff's legal counsel. The *complaint* sets forth the claims that the plaintiff (the person bringing the lawsuit) has against the defendant (the person or company being sued). The complaint generally states whether the plaintiff is seeking money damages or equitable relief, such as an injunction.

The defendant has to answer the complaint within a certain time (usually twenty days), and if he or she does not, the court may enter a *default judgment* against the defendant. The answer sets forth the portions of the complaint that the defendant admits to, if any; the allegations that the defendant contests; any defenses the defendant may have; and any claims the defendant wishes to assert against the plaintiff, another co-defendant, or any other entity who is not already a party to the litigation. These claims are known as counterclaims (against the plaintiff), cross-claims (against a co-defendant), and third-party claims (against a third party not yet part of the litigation). All of these documents—the complaint, answer, counterclaim, cross-claim, and third-party complaint—are called pleadings. In most cases, pleadings are drafted by a lawyer. However, in many courts people can file papers and represent themselves, which is called appearing *pro se*.

At this point, the defendant usually will retain or (if wise) has retained an attorney who will make an appearance before the court to submit a formal notice that the summons has been received and the defendant is represented by counsel. Then the defendant and counsel meet to determine the best course of action. A number of issues are considered, including the legitimacy of the complaint, its real value, the plaintiff's financial resources and ability to support a lengthy trial, the reputation of plaintiff's counsel, and any other issues the defendant's counsel feels are relevant.

Thus, the first stage in a typical lawsuit is the filing of initial pleadings. A lawsuit is started when the plaintiff files a complaint with the court. If a plaintiff wants to have a jury decide the case, he or she must also file a *jury demand* at this initial stage of the case.

Along with the answer, a defendant must file any *affirmative defenses* he or she may have to the plaintiff's claims. Affirmative defenses are defenses that entitle the defendant to a dismissal of the plaintiff's lawsuit, even if the plaintiff's claims are true. For example, a defendant may claim as an affirmative defense that a plaintiff has not filed his or her lawsuit within the time required under the applicable statute of limitations. If that is true, then even if what the plaintiff claims in the lawsuit is true, the case will be dismissed.

If a plaintiff does not file a jury demand with his or her complaint, a defendant can file one with his or her answer, and the case will be decided by a jury, even if the plaintiff did not want a jury to hear the case.

In some circumstances, a case may be dismissed at the initial pleadings phase, or judgment could be entered in favor of either party, if appropriate. An example of this situation is if the plaintiff fails to state a legally recognizable claim against the defendant. Depending on the circumstances, a court could give the plaintiff the opportunity to amend the initial pleadings to cure the deficiency, or it could dismiss the case altogether.

In most cases, counsels for the plaintiff and defendant will attempt to achieve settlement. If unsuccessful at first, they will likely repeat the process several times during the procedure as positions are altered by time and the course of events. If they can arrive at an agreement at this point, they will reduce it to writing and then file the document—sometimes referred to as a *stipulation*—with the clerk of the court. When this occurs, the case is recorded as closed.

Alternatively, or in conjunction with efforts to reach settlement, the defendant can enter a number of motions to test the substance of the claim. One of the most common of these is the *motion to dismiss* or *demurrer.* Through this procedure, the defendant's attorney argues that the court should dismiss the complaint because the plaintiff does not have a legal right to judgment in his favor, even if all allegations are true. Other motions can be made to challenge the court's jurisdiction over the claim or the defendant or the formal (as opposed to legal) sufficiency of the claim. For example, if the claim is vague or ambiguous, the defendant's counsel may move to require the plaintiff to file a more definitive complaint. The defendant's counsel may also move to strike from the complaint parts deemed redundant or superfluous (immaterial). These various motions create an early testing of the complaint and attitudes of the parties toward it. Each gambit changes the scenario, affecting positions relative to settlement.

Once the motions have been made and decisions reached, assuming the complaint is still active, the defendant must reveal his position with a response to the complaint. This response may take the form of a *denial,* an *affirmative defense,* a *counterclaim,* or a combination of these:

A *denial* typically admits that certain parts of the plaintiff's claims are true but denies others. This limits the areas of dispute and, if the plaintiff is unable to prove those elements that are denied, the entire case fails.

An *affirmative defense* tends to indicate that the plaintiff's allegations are true but explanatory facts have been omitted. For example, the plaintiff may allege that a geotechnical engineer failed to monitor construction and, as a result, errors in compaction were not revealed until after differential settlement occurred. The defendant would agree but would inform the court that construction monitoring was not performed because the plaintiff rejected his proposal to do so. A plaintiff's contributory negligence would be another example of facts cited in an affirmative defense, as would a claim that the statute of limitations had expired.

A *counterclaim* alleges facts that may have been asserted by the defendant had he wished to file suit. It may be based on entirely different claims and include a demand for damages far in excess of plaintiff's complaint. In essence, then, a counterclaim becomes a cross-suit and the plaintiff must respond, going through the same steps that the defendant went through in answering the plaintiff's complaint. In this instance, however, the plaintiff's answer to the counterclaim is termed a *reply.* The complaint, answer, and reply (if the answer is in the form of a counterclaim) constitute the pleadings and, once closed, they identify the only issues that can be raised at trial, with the exception of amendments to the pleadings, which can be made within certain pre-established parameters.

After pleadings are closed, either disputant can move for a judgment on the pleadings, whereby the court examines the strength and validity of the various claims. After pleadings are closed, either side files a *notice of trial* requesting the clerk of the court to put the suit on either the jury or nonjury calendar. Either side can demand and receive a jury trial. In some cases, a judge will decide on a jury trial even if the disputants prefer otherwise.

Sometime prior to the trial, a judge often will require a *pretrial conference* or *hearing,* which requires attorneys for both disputants to appear before the judge in chambers to remedy defective pleadings, eliminate superfluous issues and simplify others, agree which documents are genuine, limit the number of expert witnesses, and determine the scope of discovery. Although the pretrial conference typically is conducted on an informal basis, attending it is mandatory. Experts may also be required to attend to give the judge background information.

The judge may also order the attorneys or their clients to make pretrial *admissions*, whereby one side admits facts that help the other side exist or admits that a point being made by the other side is true. This procedure saves considerable time and cost. A judge can also order both sides to engage in *settlement negotiations* and report back at a specified time.

The parties to a dispute or lawsuit prepare their cases based on information gained through the process of discovery. Discovery consists of a variety of methods, including depositions and interrogatories.

The *deposition* is oral testimony that the opposing counsel takes prior to a trial. The deposition is one of the most important parts of a suit because more than ninety percent of cases never go to court; suits are dropped or settled as a result of the quality of depositions. Usually, this interview is conducted orally with a lawyer for the other side present and able to participate; sometimes, it is conducted using written questions. Information about a party may be secured through written interrogatories or requests to produce documents or other things. These requests may be served only upon a party to the dispute. A request for production may seek any item within a party's control. It will be helpful if the attorney gives the expert a copy of a deposition taken previously from another expert in a similar case (provided there are no restrictions against the expert's seeing this document). This will give the novice expert an idea of what the final copy of a deposition looks like.

6.3 DISCOVERY

The process of *discovery* also occurs before the trial, comprising *subpoenas duces tecum, interrogatories,* and *depositions.* The proceedings involved can be far more arduous and time consuming than the trial itself. Discovery is the process of obtaining relevant information, facts, and evidence from the other parties. Discovery allows each party to learn about and analyze facts that may (1) support or weaken its case, (2) clarify key issues, and (3) secure evidence for use at trial.

The discovery stage of a lawsuit is usually by far the most expensive segment of litigation, especially considering that the vast majority of cases are settled. Plaintiffs may cast discovery requests as broadly as possible out of fear of missing important information, an unwillingness to analyze what actually is needed for a case, or a desire to inflict expense as a means of extracting settlement. In patent litigation, discovery requests are usually targeted at two of a company's most sensitive areas: internal financial information and research and development information.

However, during the discovery stage, the expert can play an important role by advising the attorney of the type of material that will assist the case, especially at trial. Because they are not technical persons, attorneys can often miss important documents; thus, having the expert on hand or having the expert assist with drafting interrogatories can be to the attorney's advantage.

Under Federal Rule of Civil Procedure 37, a party to a dispute may bring a motion requesting that the court compel disclosure or discovery, including asking the court to order the other party to provide responses to discovery requests. Before bringing such a motion, the party must have made a good-faith effort to confer with the other party about compliance with a discovery request. A party may also request that the court sanction or punish the other party for not complying with discovery requests. In fact, procedural rules govern the discovery process and offer consequences for a party who does not cooperate with a court order directing compliance with discovery.

For cases filed in federal district courts, discovery procedures are governed by rules 26–37 of the Federal Rules of Civil Procedure. Generally, the states have similar discovery rules. Rule 26(b)(1) authorizes litigating parties to "obtain discovery regarding any matter, not privileged, which is relevant to the subject matter involved in the pending action." *Privileged information* most commonly refers to information covered by the attorney–client privilege, such as communications between the lawyer and his or her client. Disputants can discover information in a variety of ways, including: (1) interrogatories, (2) requests for production, (3) requests for admission, and (4) depositions.

Interrogatories constitute another form of discovery, whereby one party issues a list of factual questions that the other must answer under penalties of perjury. Sometimes, experts will be involved in suggesting questions to ask as well as suggesting responses. Interrogatories are used to ascertain facts, procure evidence, and secure information that will support a party's legal claims. The Federal Rule of Civil Procedure 33 governs how interrogatories are to be written and served and establishes limits on the number of questions that can be included. The responding party must answer truthfully within thirty days of being served with the interrogatories and supplement answers if more information becomes available at a later date.

Interrogatories are often sent prior to a deposition, and there appears to be no limit on the number or the nature of the questions asked. The answers can lead to further and more in-depth questions at the deposition. Answers must be provided within thirty days and under oath, although the time frame can be flexible if both sides agree; otherwise, the court can make time adjustments.

When the attorney requests it, the expert can be a great help in responding to technical questions. Usually, the unimportant questions are answered more fully than the important ones. An expert may also help the attorney construct the questions to be sent to the lawyer.

The purpose of interrogatories is to help each side learn who the opposing expert will be, the opinions of the opposing expert, and, in the main, what the opposing expert will testify to. The attorney should find out all about the opposing expert and his or her qualifications. The expert can be of help here, too, by finding out if the opposing expert is a member of a learned society or if he or she has published papers or books in this general or specific field.

Requests for production are covered under Federal Rule of Civil Procedure 34, which permits a party to a dispute to request that another party produce for inspection and copying certain categories of documents and electronically stored information. The parties involved in the dispute may also request to view tangible items, such as a defective piece of equipment that is the subject of the lawsuit, and to enter on to land or other property to inspect it, take pictures, or take samples. The responding party must answer within thirty days of the request being served and may object to the requests.

Requests for admission are covered under Federal Rule of Civil Procedure 36, which states that a party to a dispute may serve another party with a written request to admit certain facts relevant to the lawsuit or to submit proof that certain documents are genuine. If the party fails to answer or object within thirty days, the matter is deemed admitted.

At a deposition, an attorney can question witnesses, who are under oath, about their knowledge of the relevant facts of the case. A court reporter records the testimony. Attorneys can object to questions based on evidentiary rules. The person being deposed (the deponent) will still be directed to answer questions to which the attorney objects, but the attorneys will seek a ruling from the judge as to the admissibility of such answers before trial.

Depositions give the attorneys a preview of testimony that they will likely hear at trial so that there is no surprise. Depositions serve to preserve on the record the recollections of a witness while they are still fresh. In most cases, many months (and sometimes years) may elapse between the disputed event and the eventual trial. Memories of a specific event can become clouded as time passes, or they can fade away altogether. In the event a witness dies or is otherwise unavailable for trial, his or her deposition testimony may be read into the record. In addition, depositions are usually videotaped in anticipation of the witness's not being available to testify at trial.

The discovery procedure serves to expedite both trial preparation and the trial itself and requires mutual disclosure of evidence, thus permitting opposing attorneys to narrow the focus of dispute. Discovery also eliminates the needless time and expense that otherwise would be required to obtain certain facts, and it commits witnesses to the version of the facts they relate during discovery. In addition, discovery creates a better basis for attaining an out-of-court settlement, preserves for use at trial evidence that may otherwise change or disappear, and removes the element of surprise at the trial.

A *subpoena duces tecum* is issued by the clerk of the court (or in some jurisdictions by the attorney of record) to compel the other party to provide certain documents. Delivered by a process server, it usually is accompanied by an affidavit that identifies more specifically what is sought and why it is material to the case. The opposing counsel may attempt to reduce the number of documents sought by reaching an agreement with the other attorney or by seeking a protective order from the court. The expert also is subject to a subpoena duces tecum.

Depending on the particular circumstances of a case, different discovery tools may be better suited for obtaining the desired information. However, a party may use as many or as few of these discovery tools as needed to accomplish its purposes.

Many companies and organizations that may be prone to lawsuits use attorneys to determine what steps the company or organization can take to protect documents against future excessive discovery requests. This requires having a strategy prepared in advance of litigation, such as plans to preserve and protect information from discovery and to keep the litigation focused on discovery that is reasonable (to the company or organization) in terms of scope. This becomes a word game and is an area where the technical expert can be of great help in determining the wording of the request for information.

Discovery can often be very time consuming and expensive. Indeed, discovery is usually where much of a client's money is spent during litigation. The good news is that it is not uncommon for cases to settle after discovery because a party lacks sufficient facts to pursue the claim further and voluntarily resolves the case or because the court grants a motion to dismiss the case.

As the production of *electronic data* in litigation has increased in volume, so has its scope. Early forays into electronic discovery focused primarily on e-mail messages and, to a lesser extent, electronic versions of word processing documents. Attorneys, however, have become increasingly aware of new data types that may lead to useful evidence, and they are regularly issuing broader electronic discovery requests. Unfortunately, the technical understanding necessary to make informed requests has, in many cases, lagged behind the desire to do so. As a consequence, scientific and engineering technical experts are required to ensure that requests for electronic discovery are not too broad and unfocused, leading to failure to acquire relevant information from electronic databases.

Briefly, a database is a collection of data arranged for easy computer retrieval or *a collection of nonredundant data that can be shared by different application systems.* The requesting party, through the expert, should seek to determine the nature of the databases available. In many cases, understanding the basic structure of a database—its fields, query forms, and reports—can go a long way to help focus requests for information. Furthermore, in some instances, a focused and guided on-site inspection of data structure may be appropriate if the producing party cannot provide written information regarding the database structure. In addition, the production request must be specific, and information regarding field definitions, query forms, and reports should be provided to help refine the request. The data thus produced must be in a form usable to the requesting party but will vary from database to database.

Finally, the expert is well advised to remember that a party to a dispute or litigation has no duty to come forth with information before trial unless the opposing party specifically asks for it. Although rules, local practice, and judicial management orders may set outside limits on the number of interrogatories or depositions or set the deadline to complete discovery, judges are generally not involved in the day-to-day conduct of discovery, intervening only when one party objects to specific questions or requests.

Once discovery has been completed (and sometimes before), the defendant or plaintiff (or both) will typically ask the court to rule in its favor as a matter of law, without a trial. In evaluating a request for *summary disposition,* the judge will review the pleadings as well as the motions and briefs filed by each party essentially summarizing the case and articulating the reasons why that party should win the case without having to go to trial.

The judge also relies on evidence gathered by the parties during the discovery phase of the lawsuit and presented to the court in connection with the motion or motions for summary disposition. When reviewing the presented evidence, the judge must draw all inferences in favor of the party opposing the summary disposition, and he or she may grant the motion only if there are no genuine issues of material fact and the party requesting summary judgment is entitled to it as a matter of law.

If a judge grants a motion for summary disposition, the case does not proceed to trial. However, the case may not be over, because the party who loses a summary disposition motion is entitled to file an appeal of the judge's decision. However, if the judge does not grant a motion for summary disposition, the case will proceed to trial, and the next stage is the deposition phase.

More than likely, the expert (among others) will be summoned to appear at a deposition, during which he or she will be under direct examination by the attorney and cross-examination by the opposing counsel. Moreover, being called to a deposition is not just a matter of showing up and answering questions; it requires careful preparation by the expert and making sure that he or she is capable of answering the questions posed by opposing counsel.

6.4 PREDEPOSITION PREPARATION

It is far more likely that an expert will be fully involved in a *deposition,* a form of discovery whereby the opposing attorney asks questions of the other side's witnesses. Each witness is required to take an oath; a court reporter transcribes everything that is said.

Experts frequently are deposed, and they often work with counsel to develop questions for the other side's expert. In some cases, the witnesses for both sides attend depositions: to answer questions, to provide questions, and to provide guidance for other questions based on prior answers. It is not unusual for an expert's deposition to last from one day up to a full week. The transcript of a deposition can be used as evidence in court. In addition, procedural rules for depositions and other forms of discovery address a number of concerns, including

how the deposition is conducted;
the permissible scope of the deposition;
who may conduct a deposition; and
the moment when a party may object to a question at a deposition.

Before the expert's deposition is taken, a considerable amount of preparation is necessary. The expert and attorney must work together, and the technical aspects and specific facts of the case must be reviewed, even if these subjects have been covered between these parties months before; if necessary, any paralegal assistants involved in the case should also participate. In many cases the expert will be deposed by opposing counsel and will be asked to participate in the development of questions used to depose other witnesses (particularly the opposition's expert).

A deposition amounts to a cross-examination without the benefit of direct examination and in many other respects is similar to what would occur at trial (if there is one). Counsel will appear with the expert. A court reporter will be present to develop a transcript. All witnesses testify under oath. If the case does proceed to trial, the transcript is a key factor in developing questions and overall strategies.

The attorney and the opposing counsel will set up a convenient time for the deposition. Prior to this, a subpoena may be served on the expert witness, who should not be dismayed, because this is the correct procedure. Opposing counsel tends to think that such a serving may upset the expert witness. It may or it may not. The venue for the deposition is arranged by either the attorney or the opposing counsel. The expert witness should never agree to have a deposition in his or her office because of potential distractions resulting in poorly thought-out answers to the questions.

In some jurisdictions there are limits on how much notice must be given, how far away the deposition's locale may be, and other housekeeping issues. In all cases, the expert should notify counsel that the subpoena has been served and, if the dates are a conflict with other commitments, counsel should establish alternate dates.

To prepare for a deposition, the expert should work with counsel to develop the questions that he or she will likely be asked as well as the answers. In all cases, the expert's answers *must* be honest. However, it can be highly damaging to the expert's client's case to provide more information than is necessary to answer a question. For this reason, counsel should hold a predeposition conference with the expert. During this conference counsel arranges for a mock deposition, asking questions the expert has developed and perhaps even inserting trick questions to determine the effectiveness of the expert's answers and demeanor. Counsel may even ask the expert to shade a response one way or another; however, in all cases, the expert must resist pressure to provide an answer that, in his or her opinion, is false or misleading.

It also is common for experts to assist their attorneys in developing hypothetical questions that will be asked of the opposition's expert or opposing plaintiff or defendant peer. The expert may even work at developing follow-up questions and serve as the mock deposee to evaluate and help improve his or her attorney's performance.

The expert should be careful to delineate exactly what type of deposition work will be required. This could include the expert's presence at the deposition of the opposing expert or opposing peer to develop follow-up questions for counsel to pursue. In fact, if extensive deposition activity will be necessary, the expert must face the realization that several weeks or more of work are required.

As part of the preparation, it is also worthwhile for the expert and attorney to review and confirm the dates of events. The opposing counsel may rightfully ask questions related to the initial contact of the attorney with the would-be expert or related to the date that the expert formed an opinion and the circumstances that influenced that opinion. The opposing counsel is trying to determine if the attorney played any role, such as prompting or assisting the expert to form his or her opinion. Opposing counsel may also ask other questions to determine if the expert is a professional expert witness.

Although the time and place of the deposition have been discussed earlier, the expert should never agree to a deposition taking place in his or her office. The distractions are too great for the expert to be effective and concentrate on the matter at hand: the deposition.

At the deposition, the attorney and the opposing counsel will be present, along with all other attorneys who may be involved in the case or who may otherwise have a professional interest. Also required to be present will be a court reporter, who should be a certified shorthand reporter as well as a notary public. This individual will administer the oath and then make a verbatim official record of all the questions and answers at the proceedings.

There are times when, for some reason or other, it is not possible to assemble for a deposition. Under these conditions, a deposition may be taken and recorded on videotape. This videotape can be played before the jury. In recent depositions, videotaping of the expert during the deposition has become almost standard procedure. Both the attorney and the opposing counsel have the right to receive copies of the videotape. They will review it to determine which question made the expert (literally) squirm in his or her seat. If the expert has any disability, such as a back injury or leg injury that causes discomfort when sitting in a seat for several hours, the attorney should introduce this (by whatever means he or she chooses) at the beginning of the deposition.

All expert witnesses must be made aware that failure to appear for a scheduled deposition can be costly. It is possible for a judge to hold the forgetful expert in contempt of court or levy a fine on or even incarcerate him or her. The judge has the prerogative of then barring the expert from testifying at the actual trial.

6.5 ITEMS FOR THE DEPOSITION

The expert is expected to take all documents previously requested by the opposing counsel, sometimes even handwritten notes, to the deposition. There may even be a request for the expert to take transcripts of previous depositions he or she has given as well as copies of his or her testimony in other trial cases. All of these may become exhibits for the trial. The attorney should not only be aware of but also see all such documents before the deposition is taken. If it is not possible to show the attorney all documents before the deposition, it is best, when asked for the publications, to hand these to the attorney who, in turn, will glance at each document and then will give each to the opposing counsel.

At times the expert may rely on a special reference to give a correct answer to a question. That reference may already be marked as an exhibit; if not, the opposing counsel may (or may not) wish to give it an exhibit number. If the reference is not marked as an exhibit, the attorney may make a note to be sure that it is entered into evidence at the trial. The expert should exercise judgment about the use of references at the deposition (and even at the trial). If the opposing counsel reads a quotation to the expert, it is proper (even mandatory) for the expert to request to see the original document before commenting. Use of references during a deposition or trial must first be discussed with the attorney.

The question of the expert's taking notes at a deposition should be thoroughly discussed with the attorney as part of the predeposition preparation. Taking notes can be misunderstood; opposing counsel may inquire about the reason for taking notes, and the notes may also appear at trial as an exhibit. One disadvantage to taking notes is that the attorney may not have a chance to see what the expert has written and thus study its effect on the case. Another disadvantage is that the expert's notes may give the opposing counsel new ideas and hence new approaches to the problem. There is also a chance that if the expert takes notes, the opposing counsel can use words in the notes to try to undermine the expert's credibility as a witness.

In summary, showing up at the deposition with several references and the possibility of taking notes are not always good ideas and can even be detrimental to the case.

6.6 THE DEPOSITION

This procedure is similar to a court appearance and has the same legal standing and offers opposing counsel the opportunity to examine the expert witness, who must remember that at all times he or she is under oath. As a rule, in the absence of a jury as an audience, some of the usual pyrotechnics and posturing of the various attorneys will not take place. However, this does not mean that opposing counsel will not ask hostile questions or harass the expert witness.

The attorney and opposing counsel will be present, along with all other attorneys who have a stake in the outcome of the case or who may otherwise have a professional interest. Also required to be present will be a court reporter, who should be a certified shorthand reporter as well as a notary public. This individual will administer the oath and then make a verbatim official record of all the questions and answers at the proceedings. If no court reporter can be present, the attorneys may elect to have an informal deposition taken. In this case, the opposing attorney will question the expert, but no official records will be taken and the notes will not stand up in court. The expert witness should be cautious of such informal depositions because they merely serve to allow opposing counsel to get a measure of the expert and study his or her reactions to various types of questioning.

The deposition is oral testimony of the expert witness that is taken by opposing counsel prior to a trial. The deposition is one of the most important parts of a suit because more than ninety percent of cases never go to court; suits are dropped or settled as a result of the quality of depositions. It will be helpful if the attorney gives the expert a copy of a deposition taken previously from another expert in a similar case (provided there are no restrictions against the expert's seeing this document). This will give the novice expert an idea of what the final transcript of a deposition looks like.

The main purpose of a deposition is to determine if enough evidence has been generated to settle a case without going to trial. In fact, the deposition may actually be a fishing expedition insofar as opposing counsel attempts

to discover what the expert knows about the case or the facts of the case, with whom he or she has discussed this case, what tests he or she has made, and his or her opinion;
to discover the science of the case;
to evaluate the expert as a good or poor witness to oppose during a court trial; and
to develop tactics later to discredit the expert as a witness or, if that is not possible, to make the expert less effective before the court.

By the time of the deposition, the expert witness should be well acquainted with and have a thorough understanding of the facts relevant to the case and how the particular controversy developed. Knowing the facts of the case is essential for the expert witness. Facts can be gathered from a variety of sources, including pleadings (e.g., the complaint or answer), documents received via discovery, and independent sources (e.g., online public records, news, and private investigators). Although the expert witness should be prepared on the law of the case, he or she should not look, act, or sound like a lawyer; however, he or she does need to know about the law applicable to the case and how the facts tie into the law. The expert should have also prepared physically for the deposition. The witness should get a good night's sleep prior to the deposition and should not be distracted by other matters leading up to the deposition.

On the day of the deposition, the witness should be prepared to give good testimony. This requires practice. Some experts make better witnesses than others; however, every witness can be prepared in such a way to be as good a witness as his or her personality and intellect will allow. This involves understanding (through the advice of the attorney) the way to dress for the deposition, the way to sit, the way to respond, and how to react when the expert becomes frustrated or irritated with the other lawyer. Remaining in control of his or her emotions is imperative for the expert, and the witness must also be prepared to deal with leading or argumentative questions.

It is essential for the expert to be truthful at all times. There should be no partial truths. The oath that the expert will take at the commencement of the deposition specifies the whole truth.

The deposition is conducted by the opposing counsel because he or she has the right to examine the expert. The procedure is similar to a court appearance and has the same legal standing; the expert must appreciate that at all times he or she is under oath. As a rule, in the absence of a jury as an audience, some of the usual posturing may not take place. This does not mean that the opposing counsel will not ask hostile questions or attempt to harass the expert.

The procedure used in a deposition may vary, but the opposing counsel starts the proceedings by introducing himself or herself, followed by giving explicit instructions about what is to take place.

6.6.1 Deposition Protocols

Seated around the table will be all the various attorneys whose clients have an interest in the case; opposing counsel may include as many as possible in an attempt to intimidate the expert. Also present will be the certified court reporter. The expert witness usually sits at the head of the table, and the court reporter sits next to the expert or, if he or she is taping the proceedings, at the opposite end of the table in order to play nursemaid to the video equipment. Also sitting near the expert will be both the opposing counsel and the attorney.

At the commencement of the deposition, the opposing counsel will ask several boilerplate questions, such as the expert's name and the number of times that he or she has given a deposition. At this time the basic rules for answering questions will be explained. The expert will also be asked if he or she has seen the notice of the deposition. Usually, the opposing counsel will spend time on the expert's qualifications. There may be questions about experience, which the expert should try to anticipate before the day of the deposition. The attorney should have given the opposing counsel a copy of the expert's curriculum vitae to review before the deposition. In all of this, opposing counsel (who has read any report the expert has submitted) is on the lookout for any information that he or she can use as a cause for disqualification of the expert.

During the course of the deposition, opposing counsel will attempt to determine if there is sufficient evidence to substantiate the case or, often, if there is sufficient evidence to settle the case without going to trial. In other words, the deposition is often a fishing expedition to determine what the expert knows about the case or the facts of the case, with whom he or she has discussed the case, what tests he or she has made, and his or her opinion. Opposing counsel is also attempting to learn much more about the science or engineering aspects of the case in terms of the valuable information he or she can glean from the expert. In more mundane terms, opposing counsel also uses the deposition to size the expert up as a good or poor witness to oppose during a court trial and, as a result, he or she will formulate tactics (1) to discredit the expert as a witness, or, if that is not possible, (2) to make the expert less effective before the court. On the other hand, the deposition also presents the expert with a chance to assess the methods used by the opposing counsel and to determine if opposing counsel really sharpens his or her teeth every morning!

The expert must be aware that anything he or she says at the deposition will, in all probability, be used at the trial. Answers to questions put by opposing counsel must be understood and the answer formulated carefully. The expert should speak calmly and answer all questions in complete sentences. By the same token, if opposing counsel cuts off an expert who is replying to a question, the expert should state that he or she had not completed the answer. If opposing counsel disregards this comment, the attorney should be prepared to ask the expert to present the answer at a later time in the deposition.

If a question is not clear, the expert should ask opposing counsel to repeat the question. If it is still not clear, the expert should repeat the question in his or her own words and ask the opposing counsel if that is the question to be answered.

In these days of multiple qualifications, opposing counsel might have a first degree in science or engineering. If so, the attorney should have advised the expert of this. Sometimes the opposing counsel will try to intimidate the expert by attempting to show that he or she knows more about science or engineering than the expert. The expert should avoid falling into the trap of who knows more. This will only add fuel to the fire of opposing counsel's questions because he or she knows that the expert can be intimidated by such an approach.

Opposing counsel's questions will be well thought out, and part of the deposition may involve a series of questions that require a simple *yes* or *no* answer. The expert should not be fooled by this because the next question may be more pertinent and not require a rhythmic *yes* or *no* answer. Although it is preferable to use complete and understandable sentences for answers, the answers should be brief and contain the points that the expert wishes to get on the record. At times, it may be necessary to elaborate on an answer, but the expert should be cautious about volunteering information that is not requested; this will only give opposing counsel new avenues to explore. The expert should also make it clear that he or she cannot answer questions related to subjects outside the expert's field of expertise.

The worst types of questions are those for which opposing counsel insists on a *yes* or *no* answer. These are often trick questions, and it is acceptable to pause before attempting to answer. The expert may also answer, "Yes, but this answer needs elaboration." Thus, the expert goes on record as showing that a simple *yes* or *no* does not give the complete story. During the trial, the judge may ask the expert to elaborate.

The expert should not guess an answer. If the expert does not know an answer, he or she should not be goaded into making a statement for the simple reason of speculation or filling the vacuum of silence. Opposing counsel may even invite an educated guess, but this can be a trap to be used against the expert at the trial. The expert should not explain thoughts on how he or she arrived at an answer, unless asked specifically. Above all, the expert should not allow opposing counsel to put words in his or her mouth. There is no *off the record* for the testifying expert.

The attorney may not speak during the deposition except to object to a question (for the record). If an objection is made, the expert should stop talking at once and let the attorneys deal with the objection. Because the objection goes on the record, the expert may, if instructed, attempt to answer the question. The attorney may be making the objection to alert the expert to a potential problem. During a deposition, the attorney cannot officially advise the expert not to answer a question.

The deposition can go on for a few hours or a few days. The expert witness has the right to ask for a break at any time. However, during a break a witness may or may not speak to anyone. Often, after a break, the opposing counsel will ask the expert if he or she spoke to anyone during the break. The expert should not equivocate but answer truthfully. If the answer is yes, opposing counsel will ask what was discussed. The objective here is for the opposing counsel to see if the expert has been coached, and the style of the answer may be more important than its substance.

For the most part, the break gives the expert a chance to think over answers to previous questions and may help him or her decide if a correction should be made. If that is the case, the expert should go on record just after the break and make that correction. If possible, the expert should explain why he or she made the correction before the opposing counsel asks; every expert makes some mistakes during a deposition.

6.6.2 The Expert's Conduct

Experts usually are not deposed until three or four months before a trial because attorneys do not want to claim a technical consultant is an expert until they are confident his or her testimony will support their client's position. Prior to the time a technical consultant is named as the expert, his or her file may contain both positive and negative information. The file is considered privileged, however. Once he or she is named the expert, only positive information should remain because the file at that time is subject to a subpoena duces tecum.

Conduct during a deposition should be essentially the same as that during a trial, even though the setting is far more relaxed. In fact, because the setting *is* more relaxed, composure and a good night's sleep before the deposition are all the more important. Every off-the-cuff remark, joke, or jest will be recorded. Something the expert says at deposition that causes the opposing attorney to laugh may be brought up during the trial, with the same attorney reading it from the deposition transcript with deadly seriousness. Likewise, despite the relaxed setting, the expert should not be relaxed about listening to the questions asked or about framing the answers. Very often the questions will be somewhat loose, in the hopes the expert will fill in certain blanks and somehow volunteer important information. If a question is vague, the expert should ask for more specificity and, in general, be alert to the various techniques that may be applied under cross-examination. The expert should be particularly aware of sentences or questions that begin with the phrase "would you agree" and that contain double negatives. Two negatives might make a positive—for opposing counsel.

The expert needs to realize that a common tactic is for opposing counsel to try to coach the deponent through speaking objections (i.e., objections in which the attorney does more than simply state the basic objection). In general, such coaching is improper and should evoke objections from the attorney. However, during a deposition, the expert can ask to confer with his or her attorney at any time. Accordingly, if the expert is unsure as to how to answer a question, he or she should ask for such a conference, being sure to ask questions out of earshot of everyone except those serving with then he or she. The expert should not rely overly on such conferences because the expert will appear to be unprepared and give the impression that he or she is not independent. Such actions are there for all present to see and for the court to examine at a later date; the deposition is being committed to a transcript and is more than likely being video recorded.

As a final word, the deposition should have provided the expert witness with the opportunity to learn about opposing counsel's case. The questions posed to the expert can inform him or her about opposing counsel's strategy in the case and the documents that he or she believes to be important as well as the issues that he or she believes are critical to the case.

6.6.3 The Postdeposition

At some later time, the expert will receive a copy of the deposition transcript. It is crucial that the expert read the transcript carefully in light of any new knowledge. Corrections should be made on a separate page, and the list of corrections may have to be notarized. Any correction (except spelling) that the expert makes will be brought up in the courtroom during the trial; the objective is to help nullify the expert's testimony.

6.7 CONFLICT OF INTEREST

A conflict of interest is a situation in which an expert has competing professional or personal interests. Such competing interests can make it difficult for the expert to fulfill his or her duties impartially. A conflict of interest exists even if no unethical or improper act results from it. A conflict of interest can create an appearance of impropriety that can undermine the credibility of the expert, leading to possible dismissal from the case by the judge.

The first step in any litigation engagement is to do a conflict of interest check. Obviously, if one of the opposing parties is or has been a client, the consultant-turned-expert will have a conflict. Prudence suggests the conflict analysis should go beyond the actual or potential parties to the lawsuit. This includes the expert's being comfortable with the position the lawyer wants the expert to take. The expert must also review pervious court appearances to ensure that he or she has not taken a contrary position in previous litigation or in a publication.

Thus, before accepting employment as an expert or expert witness, the scientist or engineer must always ensure that no conflict of interest exists. To do this, the would-be expert should obtain instructions from the attorney as to confidentiality and other client concerns. When potential conflicts exist, the scientist or engineer should seek permission in writing to consult from prior confidants. The penalties for acting as an expert when a conflict of interest exists include: (1) the expert will be disqualified from testifying, and (2) the attorney and the expert may both be disqualified from the case.

In the expert witness system, the duty of loyalty owed to the court prohibits the expert from representing anything other than the truth; there are no exceptions to this rule. A conflict of interest can never be waived, and judges take a dim view of such situations. An undisclosed representation involving a conflict of interest can subject an expert to the denial of legal fees, dismissal from the case, or, in some cases (such as the failure to make mandatory disclosure), charges of perjury.

Generally, scientific and engineering codes of ethics (Chapter 1) forbid conflicts of interest. Often, however, the specifics can be controversial. Codes of ethics help to minimize problems with conflicts of interest because they can spell out the extent to which such conflicts should be avoided and what the parties should do when such conflicts are permitted by a code of ethics. Thus, expert witnesses (in fact, all professionals) cannot claim that they were unaware that their improper behavior was unethical. The threat of disciplinary action (for example, a lawyer's being disbarred) helps to minimize unacceptable conflicts or improper acts when a conflict is unavoidable.

Previous publications can be the bane of the expert's life. In them, he or she has possibly made many statements that are contradictory to the testimony given in a deposition and then in court—except that previous testimony was made under oath. Judges and jury members recognize that scientific and engineering theories and technologies evolve and that they change as they do so. The expert can explain this and it will usually be acceptable, providing the expert is not trying to escape from a statement he or she made merely months ago.

Like previous publications, previous testimony can also be the bane of the expert's life. He or she has possibly made many statements that are contradictory to the testimony given in a deposition and then in court. Again, judges and jury members recognize that scientific and engineering theories and technologies evolve and that they change as they do so. If such changes cause a change in testimony, the expert can explain this and it will usually be acceptable to the judge and the jury.

7 The Trial

7.1 INTRODUCTION

A trial is an expensive process for resolution of disputes. An adjudicator (the judge) must consider the actual facts and then apply the law to reach a legal resolution of that dispute. The jury then makes the decision as to which party is right. Attempts are being made to prevent or minimize the expense and backlog of court cases within the judicial system. One means is by arbitration. The rules on arbitration vary considerably, and expert witnesses are often called upon to speak at arbitration hearings.

Furthermore, at some stage prior to the trial, a *motion for summary judgment* may be made by either party to obtain a dismissal of the claim or the counterclaim or both. Unlike a motion to dismiss, a motion for summary judgment alleges that the case has no merit because the disputant bringing it to the court cannot prove that the alleged facts are true. The expert may be called to assist in formulating the words for the motion or, if retained by the defendant, for the opposing counsel. Assuming that neither approach is successful, the expert must be fully prepared to give testimony at the trial.

The expert about to testify must be prepared for a long day and get a good night's rest as well as eat a sustaining breakfast, being careful not to induce gastrointestinal problems. When a case goes on to a trial, it can be tried before a judge alone (*bench trial*) or before a jury (*jury trial*). In either case, the expert witness must be prepared to be present at the time designated and perform the functions demanded of him or her.

Occasionally, just before the trial date is reached, for some reason or other a trial is postponed. The attorney must notify the expert of the postponement and when to expect the new trial. On the other hand, the expert should always call the attorney several days before a trial to be sure the date is still firm. Knowing a trial is impending, the expert must be flexible as to dates because the court rather than the attorney determines the calendar for the rescheduled trial. At this point the expert is well advised to ask if the attorney needs another meeting to discuss any additional information.

7.2 PRETRIAL PREPARATION

As the time for trial arrives, attorneys will answer the *calendar call* and the suit will be assigned a courtroom. At this time, the expert should be considering a number of issues that need to be resolved before trial. Of particular concern are those related to timing. In essence, the expert must have a good estimate of how much time certain tasks will take and when certain events are scheduled.

7.2.1 Time and Place

The expert should find out where the trial will be held—in which building, on what floor, in what wing of the building, and in what room. The expert should also determine where he or she is to meet the attorney and at what time.

If the expert is not familiar with the hearing site, he or she should ask for directions and determine the mode of transport. A five- to ten-minute walk (usually from the hotel) is ideal, provided the expert is not carrying heavy bags or boxes. If the expert is traveling by automobile, determining the approximate time from the starting point is necessary. Also, rush hour traffic problems must be taken into account. A dry run or two during rush hour will help to determine the time needed. Even then, last-minute decisions by local authorities to perform road maintenance can occur. And then there is parking! The expert needs to determine where he or she should park and the ease of accessibility of the parking area. This brings to mind the needs for parking permits and the like.

If the site is distant, the expert should be prepared to arrive the day (not the night) before the trial commences or whenever the attorney suggests. If arriving the day before trial, the expert should consult with the attorney about preferential and comfortable places to stay.

If the courtroom is local, many experts brush off the idea of a pretrial look-see or walk-about. It is a wise course of action to visit the courthouse several days before a scheduled appearance so that the expert can become familiar with the route, potential traffic congestion points, parking facilities, and the precise location of the courtroom. In addition, it may be of value for the expert to visit the courtroom to get a sense of the place before entering during the trial.

If the expert is local and believes that nervousness may be a problem, he or she should take a cab. Bus transportation may also be used, but taxis are more reliable when traffic congestion occurs. For a consideration, the driver will often go through the back routes to deliver the expert on time. The bus cannot do this.

Although such details may seem somewhat trivial, they can become all important if the expert has never testified at trial before. The expert will be nervous, and having necessary details written down can make a nerve-racking experience far easier. Above all, the expert will do well to remember Murphy's law: *Whatever can go wrong will go wrong.*

7.2.2 The Beginning and Summary

If a jury trial is called for, the opposing attorneys and judge conduct *voir dire,* a procedure whereby prospective jurors are questioned to determine their qualifications. If a juror's answer indicates a prejudice, a relationship with one of the parties, a financial interest in the outcome, or any other situation meriting disqualification, an attorney may *challenge for cause* and state his or her reason. Each attorney also is given a specific number of *peremptory challenges,* which gives him or her the freedom to excuse a juror without citing a reason.

Once the jury is impaneled (often with alternates), the trial begins. The plaintiff's attorney typically starts by sketching the facts he intends to prove, followed by defendant's counsel, who does much the same thing. Then the plaintiff calls his witnesses, each being bound by the various *rules of evidence* (Chapter 3), including the *parole evidence rule,* the *relevancy rule,* the *hearsay evidence rule,* and the *best-evidence rule.*

The best-evidence rule (Chapter 3) holds that the best possible form of evidence must be produced at trial. For example, if a document is evidence, the original (if available) rather than a photocopy would be used. It also is a fundamental rule of evidence that all nontechnical witnesses may testify only to matters of fact; they may not express their conclusions. Conclusions are to be made only by the trier of fact: the judge or jury.

By contrast, expert witnesses are permitted to state their opinions and conclusions, which become evidence, providing that (reference must be made to the prevailing statutes): (1) the matter on which the expert is testifying is personally known to him or her or has been made known to him or her at or before the trial or hearing during which his or her opinion is expressed, (2) the matter on which the expert relies to form his or her opinions is of the same type that any expert in the same field would rely on to form his or her opinions on the subject involved, and (3) the expert does not rely on any matter that, by law, he or she is forbidden to rely on (the matter relied on does not have to be admissible as evidence).

The opposing attorney can file a number of motions while witnesses are being questioned, including an objection to a question, an answer, or both. If the judge considers an objection valid, the jury (if one is being used) will be instructed to ignore the point of the objection.

Once the plaintiff presents his evidence and all his witnesses have been fully questioned, he rests. At this point the defendant can decide whether to present his case or to settle. If he decides to proceed, one or several strategic motions can be made. One of the most common is a *motion for a directed verdict.* This alleges that the plaintiff has failed to prove the case; the judge will grant a directed verdict if he or she believes all the evidence presented, even if true, would result in an unbiased jury finding for the defense. If the motion is granted, the plaintiff may not start another case on the same grounds.

If it appears likely to the plaintiff that a motion for a directed verdict would be granted, he or she reacts before resting the case by making a *motion for a voluntary nonsuit.* This would give the plaintiff the right to begin another action on the same grounds after paying court costs. It might be anticipated that the second action would be stronger than the first, filling in the gaps in the evidence that weakened the initial presentation. A motion for a voluntary nonsuit can be made any time before the judge renders a decision or gives the case to a jury.

Assuming the plaintiff does not move for a voluntary nonsuit and that a motion for a directed verdict is denied, the defense presents its case, following the same procedures applicable to the plaintiff. After presenting the last witness, the defendant may ask for a motion for directed verdict, whether or not this was sought before. If the motion is denied, the plaintiff may offer evidence to *rebut* what was said by the defendant's witnesses, and the defendant may subsequently introduce evidence relative to that introduced during rebuttal.

Once all the evidence has been heard and both sides rest, either or both can ask for a motion for a directed verdict. If none is granted, the plaintiff presents a *summation,* recapitulating the claims, commenting on the evidence, and stating the legal principles involved. The defense attorney does likewise, with the plaintiff's attorney being given an opportunity for rebuttal.

After the lawyers have spoken, the judge *charges the jury*—reviewing closing arguments, pointing out the most important issues of law, and summarizing the testimony and how the jury should evaluate it. He usually points out that whoever brought the claim or counterclaim has the *burden of proof,* with proof being based on a *preponderance of the evidence.* Either or both attorneys may also present *requests to charge,* either orally or in writing, outlining special charges for the judge to consider. Either attorney can also object to the charge as given.

After it is charged, the jury goes into the jury room to discuss the evidence. In some jurisdictions, the jury must reach a unanimous agreement in order to render a verdict. In others, a majority agreement is all that is needed. If an agreement cannot be reached, a *hung jury* results, and the case must be retried.

After the jury delivers its verdict, the attorney for the losing side may make a motion for *judgment notwithstanding the verdict,* or *judgment NOV.* This is essentially the same as asking for a directed verdict, except at this point it would require the judge to reverse the jury's decision. A judge may be more inclined to grant a judgment NOV rather than a directed verdict, however, due to the appellate laws. In essence, if the judge errs in granting a judgment NOV, no retrial would be necessary; the jury's verdict would be restored.

If the motion for judgment NOV is denied, the losing party may then make a *motion for a new trial.* Such a motion could be granted for reasons similar to those that would result in a favorable ruling on a motion for judgment NOV or for reasons such as excessive damages being awarded, a procedural error on the court's part, or the opposing party's use of a surprise that could not have been guarded against.

If the judge's or jury's verdict is not set aside and no new trial is granted, the judge directs entry of a *final judgment* for the successful party.

7.3 TRIAL PREPARATION

Preparation for trial by the expert involves reviewing a number of issues with the attorney so that the expert knows what is expected. Such apparently mundane issues as the time of arrival and where to sit can cause embarrassment at the beginning of a trial because the expert was unaware of these details. Once embarrassment is evident, opposing counsel will have his or her teeth sharpened for the attack.

An issue that is often ignored and not discussed by the expert and counsel is *on-call service,* which means that the expert must set aside time during which he or she may be called to testify, depending on the court calendar. This usually appears to be inconvenient insofar as it may restrict any of the expert's travel plans or cause him or her to leave a meeting at an inappropriate moment. However, a conscientious attorney will seek to minimize on-call requirements by keeping the expert posted on developments. In some cases, the attorney may demand that a large block of time be set aside by the expert. Chaos theory intimates that no matter how well the expert prepares, the unexpected is inevitable; thus, the best way to help minimize unexpected calls is for the expert to be available.

In terms of on-call availability, the expert must recognize that even relatively simple issues can result in a lengthy trial because most trials are frequently interrupted by numerous other matters to which a judge typically has to tend. An experienced attorney usually can give the expert a reasonably accurate estimate of the number of days during which the expert will be involved, and he or she should also inform the expert of the parts of the trial that he or she should attend. Some attorneys prefer that the expert be on hand until the case is given to the trier of fact to decide. The expert's schedule must be kept open at least for the days likely to be involved and, very often, for at least one or two days after the anticipated close of the case.

Trial preparation may go beyond preparation and rehearsal of testimony and development of whatever graphics, models, or demonstrations the expert may need to support testimony. For example, the expert may be called upon to help develop direct-examination and cross-examination testimony of others on the same side, as well as hypothetical cross-examination questions counsel may use on the opposing expert or opposing peer.

Even before the trial, all materials relative to the case must be properly stored. For documents, this may include a fireproof file. For evidence or samples, however, far different storage may be appropriate in order to help ensure their security and integrity.

Finally, as the day of the trial nears, it is likely that the attorney will contact the expert to review a number of issues at different times during the course of the case's maturation.

7.3.1 Preparation of Testimony

The expert may be called upon to provide two types of testimony; the more common is that which the expert will give orally at trial under certain common guidelines relative to direct examination. Also, prepared testimony can be inserted or read into the record. If a final report has been prepared, it could feasibly be used as this second type of testimony.

The testimony the expert will give orally starts with direct examination, in which counsel may not ask the expert leading questions but can pose hypothetical questions based upon the facts in the case. Although the expert will have done some of the preparation for his or her deposition and it will be of value, there is a better than average chance that he or she will need to prepare for a number of new hypothetical questions for the trial. They must be designed and interrelated in a manner to ensure that facts, opinions, and conclusions are presented in a logical progression so that the trier of fact better understands the scientific or engineering issues involved.

For this reason, oral testimony should rely to as great an extent as possible on common English usages, as opposed to technical terminology and buzzwords. The highly technical aspects of testimony generally can be confined to formal prepared testimony inserted in the record or can be brought out through cross-examination.

During direct testimony, the goal is to acquaint the trier of fact with the expert's opinions and conclusions, which become evidence. The expert should avoid the common mistake of dwelling on the scientific work that leads the expert to his or her opinions and conclusions. Although such work may be fascinating to the expert and other technically inclined individuals, the judge and the jurors usually proceed under the assumption that the expert knows what to do to reach conclusions, unless an unusual test of some kind was performed. When commonly accepted research methods are employed, they need to be touched upon only lightly in oral testimony; elaboration can be provided in written testimony. This approach permits the expert to concentrate on the opinions and conclusions. By using common terms supported by analogies to everyday events or conditions—as well as graphics, models, and demonstrations, as appropriate—the expert will help ensure the understanding required to make his or her testimony as effective as possible.

Preparing oral testimony for direct examination should always involve a thorough review of all applicable deposition transcripts and should always include review by counsel. This review will lead to refinement, and it is not uncommon for experts to develop several drafts before a final version ultimately is derived. In all cases, it also is advisable to rehearse oral testimony with counsel at least once to help ensure that both the expert and the attorney are completely familiar with what will be asked and how it will be answered. Time also should be allowed for dry runs with others, at home or at the expert's office. The more familiar the expert is with the points that he or she wishes to make, the more convincing he or she will be.

The expert should bear in mind that the case may be decided entirely on the transcript. Although half-sentences may be fully understood within the context of the trial, they may convey the wrong impression or even the wrong answer when recorded word for word by the court reporter. Likewise, if the expert will be using graphics, he or she should be very explicit about the content of each slide or chart without any form of prompting; not to do so only serves to weaken the expert's credibility. Taking notes to the witness chair is an option, but the expert needs to be aware that the judge and opposing counsel can examine whatever the expert takes to the witness chair. The expert should be cautious about any such notes, and it should only be done with approval of counsel. In addition, the judge and jurors know that anyone can read from a prepared text. However, the expert should be able to relate facts, opinions, and conclusions off the top of his or her head—especially if he or she is truly familiar with the case involved.

The expert's testimony will also include answers to questions posed during cross-examination. The expert should be prepared for this cross-examination by virtue of what he or she has done for deposition preparation. Nonetheless, different facts may have come to light since the time of the deposition (which could have occurred months or years before) because new or altered opinions may exist. Also, a review of deposition transcripts may indicate a number of weak spots (some of which usually will be anticipated through inclusion in direct examination) or certain positions that opposing counsel is trying to establish.

The manner in which opposing counsel pursues a cross-examination in court usually will differ markedly from the approach taken during a deposition, in that much more of an effort may be made to attack the expert's credibility, capabilities, research efforts, and any other factors that opposing counsel can bring to bear. The expert cannot know for a certainty to what extent, if at all, opposing counsel may take this line of attack. However, valuable guidance can be obtained through review of transcripts of similar trials where opposing counsel has served. Often, a pattern will emerge in which it becomes obvious that the same questions are asked time and time again, and the expert can also see the types of responses that seem to be effective or counterproductive.

7.3.2 The Expert and Direct Testimony

The initial procedure is basically the same for most trials. The expert witness is called to the witness stand and the bailiff or responsible court officer administers the oath that requires the expert to affirm that he or she will tell the truth. At all times, the expert's body language is important because it is a subtle form of communication. It tells a lot to the opposing counsel. He or she will note how the expert answers or refrains from answering a question. Hand movements, facial expressions, body movements, and sitting posture are all telltale signs of the expert's degree of comfort or discomfort.

It is essential for the expert to remain calm and collected. Although he or she is in a legal domain and may feel out of place, the expert must remember that he or she is one of the world's leading experts in the area of science and engineering pertaining to the case. As such, both the attorney and opposing counsel are entering into the expert's domain. As these thoughts pass through the expert's mind, he or she should sit down in the witness box and take a few seconds to adjust the seat and the microphone to a comfortable height and position. Comfort is of the essence because the expert may be on the witness stand for several hours.

The attorney now takes over and commences the direct examination of the expert witness. This will include questions about the expert's background, and the answers allow the judge to decide if the expert really qualifies as an expert for the case at hand. The initial questions are almost pro forma. The questions focus on name, address, occupation, and employment details. It is permissible to spell out any difficult words for the court reporter, and this can continue throughout the expert's testimony. Court reporters use a phonetically based system that may produce add spellings of names and technical words if not presented correctly to the reporter. Some attorneys present the court reporter with a list of spellings of technical words the expert is likely to use.

Then will follow questions about the expert's educational background. The qualification continues, with the attorney asking the expert about his or her experience since graduating, membership in professional associations, offices held, research conducted, scientific papers published, and the like. It is also beneficial to mention papers published in peer-reviewed journals, with emphasis on the reasons for peer review. This is especially helpful if the expert has published several papers related to the topic of the present court case; this can be addressed as the expert looks directly at the jury.

At all times, the expert should remember that the court reporter is taking all of the answers and statements down verbatim; the key is for the expert to speak slowly and clearly. Many scientists and engineers get so involved in their testimony that their rate of speech may be too fast for the court reporter. As a result, many important items are omitted from the transcripts of the trial.

It is important for the expert witness to appear properly qualified for the jury to be duly impressed. The expert should recall that his or her answer to each question and the time taken for the answers is time well spent. The questions posed to the expert are for the benefit of the judge and jury. They may be the same as those asked during the deposition. As in the deposition, education and experience are the most important aspects of qualification. However, in court, bragging and false modesty will be easily detected by the judge and jury, and it is the judge who makes the determination if the expert qualifies for the case at hand.

If the expert has a PhD or DSc, he or she must be sure that the judge and jury understand the nature of these advanced degrees and that (through the attorney) the expert should be addressed as *doctor.* However, the expert or the attorney should differentiate between the expert's title and the title given to a medical doctor. If necessary, the expert may repeat information about his or her background that was used to qualify the expert as an expert witness for the case. The jury needs to understand that the term *doctor* is not specific to the medical profession.

After the scientist or engineer has been questioned by the attorney as a means of presenting his or her qualifications to the court, counsel will then request the court to qualify the scientist or engineer as an expert witness. At this point the court (the judge) will ask the opposing counsel if there are any objections. Often there are objections, and the opposing counsel will ask to take the witness on *voir dire*. Opposing counsel will ask questions of the witness, much like cross-examination but related solely to the witness's qualifications to testify as an expert.

At the conclusion of the questioning on voir dire, the opposing counsel should state the objections, along with the grounds (such as a lack of experience or education in the field), or state that he or she has no objections. If there is no further argument of counsel, the judge will then give the ruling. If the witness shows that he or she has education or experience in the field greater than that of a layman and the witness is not predisposed to be biased toward one side or the other, the witness should be recognized as an expert.

Only after the expert's qualifications are accepted by the judge and he or she is recognized as an expert witness will the formal part of the testimony begin. The attorney will usually ask the expert a number of questions about the case.

Many will be about the facts as the expert sees them, and the answers should be made with sufficient certainty to let the jury know the expert is definitive and not equivocating. In this sense, answers such as *I guess, I think,* and *I assume* should, for the period of the court appearance, be stricken from the expert's vocabulary. At the same time, the expert should avoid any discrepancies between the answers in his or her deposition and answers to questions at the trial. At this point in the trial, the attorney is building up to the point where he or she will ask if the expert has an opinion on the case. The expert's answer should be only one word, a simple *yes.* The expert should not elaborate until asked to do so. Then the attorney will make sure that the expert has sufficient time to explain the expert's answer.

Being under oath, the expert must answer all questions truthfully. When answering a question, the expert witness must never think of how colleagues might react or offer criticism of the answers. All questions should be answered briefly and to the point, using lay language whenever possible. Tendencies to give a rambling answer will only play into the hands of the opposing counsel during cross-examination; words will be twisted, and the expert may even give the answer that opposing counsel wants to hear.

If the use of technical terms is necessary, the expert should give a *short* explanation of the terms. If the attorney asks the same question twice, the expert should assume that either he or she did not make himself or herself clear or that the answer was not what the attorney thought it should be. In this case, the expert should not repeat his or her previous answer word for word but, rather, should answer the question using different language, while keeping the same thought.

However, if pretrial preparations have been conducted, the expert will have discussed the questions to be asked at the trial with the attorney and will have formulated answers prior to being on the witness stand. When making preparations, the expert, with the attorney, should decide which points need to be made; he or she should finish the answer with the most emphatic point. Juries tend to remember the last point made and favor strong, competent answers (cf. the military analogy of remembering and obeying the last order received).

The expert should always remember that the members of the jury are interested in what he or she has to tell them. The jury has to make a decision, and to do this, the members need to be informed. Again, the expert should not talk down to the jury but should provide information using simple language. Generally, jury members are uncomfortable with the expert witness who always agrees with his or her attorney, because they may feel that the expert is the attorney's foil or puppet. Providing there are no surprises, the expert and the attorney can agree during the pretrial preparation that the attorney may ask some questions that require the expert to disagree mildly. Alternatively, during the direct examination the expert should not hesitate to correct the attorney's question and also to respond *no, not exactly,* or *that is not correct* and then follow up with a brief statement of correction.

As in the deposition, any discussion that an expert witness has with the attorney during the break or during lunch is *discoverable.*

7.3.3 The Use of Visual Aids and Notes

An expert witness may refer to notes if necessary but should make proper explanatory remarks. Notes may be useful for placing exact dates and times on the record or as examples of mathematical calculations. If the expert refers to notes too often, however, the jury may wonder how much the expert really knows about the case, how much thought he or she has given to the problems of the case, how much the expert is relying on hints given to him or her, and how much real preparation the expert has made for the case.

At times, the expert may rely on books for specific information; these books must be made part of the evidence and have exhibit numbers. Opposing counsels have been known to object to an expert's referring to a standard reference book if it has not been placed in evidence beforehand.

Visual aids can be extremely valuable in supporting testimony. An individual's retention of information related orally may be low, but it may be high when considering a demonstration. In addition, many courts are favorably inclined toward graphics, models, and other devices that better explain technical matters because they assist the jury by making technical issues more understandable.

Assuming budgets allow the development of visual aids, in all cases, materials should be well developed. This does not necessarily mean they must be elaborate, but they have to demonstrate the point, which can be ascertained beforehand by trying them on someone without a technical background. If plans are involved, it usually is better to prepare a diagram rather than using the originals; the diagram can use color and other techniques to differentiate one area or one condition from another. If a graph is to be used, the expert should rely on a natural scale rather than a logarithmic scale. In all cases, large, boldface print should be used. If it is necessary to show geological formations, the expert may wish to use a three-dimensional model rather than a two-dimensional illustration.

If the expert intends to project an image using slides, overheads, or filmstrips, it is necessary to ensure that the courtroom can accommodate such plans. It must be capable of being darkened and should provide enough room for setting up a screen (where the judge and jury can easily see it) and the necessary projection equipment. The expert should take along an extra projection bulb, an extension cord, and a projection table.

In some instances a concept can be illustrated well only through a demonstration of some type. In these cases, the expert should be absolutely certain that the demonstration will work exactly as planned before the trial starts. He or she must also be certain that the attorney is aware of such plans. Some demonstrations require more than commonly found objects; in these instances, a working model should be considered. Although working models can be expensive, they can also have dramatic impact. If such a model or other type of demonstration is too large or too complex for courtroom demonstration, it usually can be videotaped. Witnesses generally are required to watch such demonstrations and, if necessary, to testify in court as to what they saw and what transpired when the videotaping or other filming equipment was shut off.

Finally, unless the expert has access to high-quality, in-house capabilities, it is preferable to have graphics and models prepared by outside specialists who often can prepare a better finished product.

The opposing counsel has a right to ask the expert to turn over to the court all notes and books used during testimony. The books certainly will become exhibits. The expert should also inform the court now that any books are to be returned after the trial is over.

7.4 COURTROOM LAYOUT

An expert witness should not walk into completely strange surroundings, but he or she may not be allowed in the courtroom during the trial before being called to the witness stand. If the expert cannot be shown the actual courtroom, the attorney should explain the physical layout of the room in detail. A diagram or sketch would be helpful. The expert witness should know where the judge sits relative to the doors. The jury box and its position relative to the judge's seat should be described. Similarly, the position of the witness box should be identified as well as the placement of a microphone for use by the witness. At this time, the expert should ask if drinking water will be available.

If the expert has any problem with the room or its layout, the attorney should be notified at once. For example, if the expert wishes to have a table next to his or her chair for a briefcase or a small exhibit, this should be arranged before the expert takes the stand. The expert may wish to have a clear view of the jury area to see each member clearly. Arrangements of exhibits and a chalkboard or whiteboard must be thought out; neither of these should obscure the expert's view of the jury.

7.5 JURY TRIAL AND BENCH TRIAL

A *jury trial* is a legal proceeding in which a jury makes a decision or makes findings that are then applied by a judge. In most common law jurisdictions, the jury is responsible for finding the facts of the case and the judge determines the law. These jurors (the *peers of the accused*) are responsible for listening to a dispute, evaluating the evidence presented, deciding on the facts, and making a decision in accordance with the rules of law and their jury instructions. Typically, the jury only judges *guilty* (or *not guilty*); the actual penalty is determined and set by the judge.

Some jurisdictions with jury trials allow the defendant to waive his or her right to a jury trial, thus leading to a bench trial. In the United States, jury trials are available in both civil and criminal cases. In Canada, jury trials are compulsory for crimes for which the maximum sentence exceeds five years and optional for crimes for which the maximum sentence exceeds two years but less than five years. However, the right to a jury trial may be waived if both the prosecution and defense agree.

In countries where jury trials are common, the jury is often considered to be an important check against the power of the state. Other common assertions about the benefits of trial by jury are that it provides a means of interjecting community norms and values into judicial proceedings and that it legitimizes the law by providing opportunities for citizens to validate criminal statutes in their application to specific trials. Many observers believe that a jury is likely to provide a more sympathetic or fairer hearing to a party who is not part of the government—or other establishment interest—than would representatives of the state.

The positive belief that jury trials are a fair means by which justice is dispensed contrasts with other beliefs in which it is considered risky for a person's fate to be put into the hands of an untrained lay-group (the jury) that is susceptible to media and public influence. A *bench trial,* according to the laws of the United States, is a trial before a judge alone in which the right to a jury trial has been waived by the necessary parties (or there was no right to a jury trial). The bench trial is to be distinguished from a jury trial, where the jurors determine guilt or innocence.

With bench trials, the judge is the finder of law and the finder of fact. In some bench trials, both sides have already stipulated to all the facts in the case (such as civil disobedience cases designed to test the constitutionality of a law). These are usually faster than jury trials due to the lower number of formalities required. For example, there is no jury selection phase, no need for sequestration, and no need for jury instructions.

A bench trial has some distinctive characteristics, but it is basically the same as a jury trial without the jury. For example, the rules of evidence and methods of objection are the same in a bench trial as in a jury trial. Bench trials, however, are frequently more informal than jury trials. Without the time-consuming showmanship of a jury trial, the disputants are often able to concentrate on the real questions at issue. It is also less necessary to protect the record with objections, and sometimes evidence is accepted *de bene,* or provisionally, subject to the possibility of being struck in the future.

7.6 THE JUDGE

Judges come from varied educational and social backgrounds, and the vast majority are not well versed in science or engineering. One can assume that just about all judges have both college and advanced law degrees. Many are very interested and will follow testimony carefully. Some will ask questions of the expert to elicit clarification. If this happens, the expert should always speak directly to the judge and always address the judge as "Your Honor." The judge may be more interested in the content of the testimony of the expert than the jury. Very seldom will a judge show off his or her knowledge.

There have been occasions when a judge has engaged an expert in a conversation remote from the case at hand so that the audience in the courtroom could appreciate the judge's knowledge of chemistry. In such cases the expert witness must play it by ear. It is important to know that in 1992 a Supreme Court decision gave the judge fairly extensive latitude in allowing or disallowing evidence to be heard by the jury.

On the other hand, some judges are uncomfortable with, or are even suspicious o,f scientists and engineers. Judges may have heard the testimony of many experts over the span of their judicial careers, and some are inclined to believe that an expert can be found to testify to the truth of almost any factual theory. It is worth remembering that an expert's testimony can be used to confuse what may be a simple case, and the judge and jury members can be misled by such experts-for-hire.

Occasionally, an expert witness will find the judge in a case to be prejudiced against the case on which the expert is expressing an opinion. This situation can become a form of harassment of the expert. The judge may prevent expert witnesses from responding to questions posed by the attorney, possibly as a result of improper qualification of the experts by the attorney.

Judges are human and can make unwise decisions, but that is not the concern of the expert; no one should argue with the judge, no matter how tempting it is. The attorneys can and should deal with this aspect; if necessary, an attorney can appeal the judge's decision to a higher court. If the judge is giving the expert a difficult time that seems more like a hostile cross-examination than genuine technical questioning, it is wise for the expert to remain polite. The jury is negatively influenced by a witness who loses his or her composure; the jury expects the expert to be polite at all times when speaking to the presiding judge.

Above all, the judge is the "boss" and has much leeway in running the courtroom and the case. He or she may have had prospective members of the jury answer special interrogatories intended to make them focus on the specific facts of the case. In doing so, the judge has hoped that the jury will understand the facts and thus be less likely to arrive at a compromise verdict. The judge wants the jury to make its decision during deliberation based on the preponderance of evidence. Unless the interrogatories are worded carefully, they may confuse the jury.

Some judges allow jurors to take notes; if so, the expert should speak much more slowly. In fact, some judges issue loose-leaf binders to jury members. Also, the judge may instruct lawyers to furnish written material to the court to be distributed to the jury members. The judge may also require that the expert file the opinion in writing prior to the trial; then, during the trial, he or she will require the expert to read that opinion verbatim. Other judges forbid members of the jury to take notes.

Perhaps one of the most important facts for an expert to keep in mind is that some judges, although they can be expected to be interested in the content of the expert's testimony, have been known to rule in favor of or against cases based on the demeanor of the witnesses rather than on the science. Some judges have accepted the testimony of the expert who is most adamant (although perhaps wrong) over that of the expert who admits that there can be limits to the science under discussion.

At the end of the testimony, the expert may hear the judge instruct the jury with the following statement: "The expert is the judge of the facts, but I am the judge of the law. It is the expert's obligation to apply the rules of the law as I define them to the facts as the expert finds them." Another judge may say, "You have heard evidence in this case from witnesses who testified as experts. The law allows experts to express opinions on subjects involving their special knowledge, training, skill, experience, or research. You shall determine what weight, if any, should be given such testimony, as with any other witness."

7.7 THE JURY

The jury looks to the expert witness as occupying a specific role: to help them decide which scientific facts are correct and which theories to believe. Some members of the jury may be influenced by the television caricature of the scientist. Often, on the screen, the scientist is portrayed as a socially inept, inarticulate stereotype of the absent-minded professor. For this reason, appearances are of paramount importance for the testifying expert witness. Some juries (like some judges) are more interested in personalities than in either the facts of the case or the interpretation of the science. Juries can be prejudiced against an expert witness who dresses in a very unconventional style. The expert's style of clothes should therefore be conservative and exclude all types of outlandish ties or socks. An expert witness has a right to wear what he or she likes or what is comfortable. However, if the expert witness wears jeans, in all likelihood the jury will not take his or her testimony too seriously.

After the expert is called, he or she should walk to the witness stand with a bearing that radiates confidence to the jury. After sitting, the expert might be well advised to acknowledge the judge and then look at the jury before the oath is administered.

Juries object to physical barriers, and the expert should avoid stacking the witness box table with papers, books, legal pads, or even sweaters. These create a physical barrier between the expert witness and the jury. If books and the like are to be used, they should be in a briefcase or box and stored on the floor or a side table near the expert. The hands should be visible, not in pockets; when speaking, the expert may use his or her hands but should avoid pointing a finger at any person.

At all times, the expert should attempt to project a relaxed appearance to the jury and attempt to disguise any special mannerisms that might be a sign of stress. Overall, the jury wants to hear an expert who can acknowledge his or her accomplishments without playing at false modesty or being overbearing.

When the expert is on the witness stand, eye contact with members of the jury is very important, although the expert should occasionally answer a question while looking at the judge. At other times the expert should look at the attorney or opposing counsel who is asking questions, without allowing opposing counsel to stare down the expert. However, it is important to remember that the only real audience in the courtroom is the jury.

When responding to the attorney or to opposing counsel, the expert must remember that the jury is looking for answers that are to the point and as nontechnical as possible. However, the expert should not talk down to or patronize the members of the jury. The expert *must* assume that the jury members are intelligent, are eager to do their civic duty, and wish to learn but are uninformed as to the scientific or engineering facts of the case. The expert must explain any complicated phenomena and the problem at hand in simple layman's language. Above all, the expert must also demonstrate the validity of the analysis he or she is presenting and the logic of the conclusions.

Some juries are suspicious of an expert witness whose practice is limited to testifying for the defense. On the other hand, an expert who testifies only for a plaintiff is, in the minds of some jury members, not suspect. It is up to the attorney to introduce this issue to the jury and lay to rest any mental objections that its members might have.

Some jury members are comfortable with an expert whose sole source of income is testifying, but other jury members might wonder if the expert has a "real" job. Jury members may take a negative view of the expert witness who remains in the courtroom after giving testimony, wondering if he or she does not have anything more important to do. In fact, the tendency for an expert to remain in the courtroom after testifying may also be suggestive of the expert's having a financial stake in the outcome of this trial. If the expert does not leave the courtroom after his or her testimony, other questions may come to the minds of jury members who will wonder whether the expert is still being paid while sitting in the courtroom or if the expert is only a professional witness.

Such thoughts in the minds of the jury may never mature if this expert leaves the courtroom as soon as he or she is excused by the judge. However, if for some reason the attorney wishes the expert to remain in the courtroom after testimony is over, the jury should be made aware that the attorney has asked the expert to remain.

7.8 THE EXPERT WITNESS IN COURT

First and foremost, it is essential that the expert get a good night's sleep before trial, especially the night before the cross-examination. Cross-examination can be a draining experience, and the cross-examiner will sometimes save the most damaging questions for late in the day, when some of the strongest points are made and won.

The expert witness will be called to the courtroom sometime after the trial has begun. In many trials, he or she must remain out of the room until called, and the attorney should have someone, perhaps a paralegal assistant, keep company with the expert, be available to make sure he or she remains on call, and be present if there is any potential interference from opposing counsel and his or her staff.

The expert witness should be told how the judge conducts the court. This includes whether or not the judge insists on short answers or whether he or she will permit a witness to ramble on. It is also important to know if the judge permits, requires, or forbids the members of the jury to take notes. More important, the expert needs to know if the judge is inclined to ask questions and whether he or she is friendly, hostile, or aloof and knowledgeable about the scientific and engineering aspects of the case. The attorney should advise the expert whether or not he or she is required to educate the judge on the technical aspects of the case at hand.

To some attorneys and for some cases, the composition of the jury as to age, gender, and ethnic, cultural, and educational background is extremely important. Furthermore, the expert needs to identify whether or not the jury members are alert or if any specific jury member seems to indicate a problem with the expert, such as general disagreement (identified through body language) with the expert's testimony.

7.8.1 Dressing for a Court Appearance

The general rule for attire is to dress in a manner that is in keeping with the image of an expert witness to the minds of the jurors. Generally, for men and women, this means wearing a dark business suit (for women, this can be a pantsuit or a suit with a skirt of appropriate length, because she may have to ascend three or four steps to enter the narrow confines of the witness box), shined shoes, and other conservative apparel, as well as removal of any adornments: no brightly colored pocket handkerchief, no fraternal pins, and no flashy jewelry. This helps ensure that the jury's principal interest is in what the expert has to say, as opposed to what the expert is wearing, and in retaining the expert's credibility.

7.8.2 Seating and Forms of Communication

In some jurisdictions all witnesses, including experts, are not allowed to be present in the courtroom while opposing testimony is being given. In others, such rules do not apply to expert witnesses; in these cases, some attorneys prefer their experts to sit with them to provide *written* or *whispered* commentary on remarks of opposing expert witnesses. The form of the commentary will be at the discretion of the attorney. He or she may prefer to pass notes to the expert based on items of the opposing expert's testimony that are not fully understood. However, some attorneys feel that having their experts sit with them is not advisable because it gives the appearance that the expert is an advocate rather than an objective and impartial professional.

The expert should be advised of or determine the lawyer's preferences for seating before the commencement of the trial or at the commencement of the trial. This will apply to seating arrangements that are to be observed both *before* and *after* the expert gives his or her testimony.

Other forms of communication come into play once the expert takes the witness stand and will be dealt with here rather than in the next section. Most attorneys advise an expert to look the questioner in the eye, but guidance on whom to look at when responding differs. Some prefer that the expert continue to look at the person who asked the question; the majority seems to agree that he or she should look at the jury, if there is one, or at the judge if no jury is present. During explanations of technical matters, it is often advisable for the expert to look at the jury and address each juror by eye contact as the expert gives the explanation. On a personal note, some experts who wear contact lenses or spectacles, such as those with trifocal lenses, may not be able to stare at one spot for prolonged periods; eye refocusing is often necessary.

Some attorneys believe that it is appropriate to develop a set of signals that permits unspoken conversation between the expert and the attorney. This might include signals for tiredness or to call a brief recess (if possible) so that the expert can discuss a recent idea with him or her. Alternatively, if, during direct examination, the attorney has forgotten to ask an important question (which is unlikely), a signal for a recess might be warranted.

However, signals are risky. Opposing counsel might object and accuse the expert witness of following the attorney's lead and parroting the evidence that the attorney wishes to bring before the court rather than the expert's unbiased opinion. Seeing such signals, the judge might choose to reprimand the attorney and the expert, thereby indicating to the jury that neither is to be trusted. Signals can cause problems in the courtroom and should not to be used because of the possible downside.

7.9 THE SCIENTIST OR ENGINEER ON THE WITNESS STAND

Insofar as a scientist or engineer's expert testimony becomes evidence and because the testimony of two opposing experts often will differ, it is up to the judge or jury to determine who is right and who is wrong.

As soon as an expert takes the stand and has recited pertinent credentials, the hiring attorney will seek to have the court recognize him or her as an expert. Opposing counsel may at this time challenge the expert's credentials. This seldom is successful and, in many instances, both opposing counsel and the judge will stipulate that the witness is fit to serve as an expert, making even direct examination of his or her credentials unnecessary. This can be detrimental in those instances where the expert's credentials are particularly noteworthy, especially in comparison to the opposing expert's credentials.

Once the scientist or engineer is recognized as an expert by the court, the disputant for whom he is appearing begins questioning through *direct examination*. Under direct examination, the attorney may not ask *leading questions*—that is, questions that show how the witness is to answer or that suggest the preferred answer.

After direct examination, the opposing attorney engages in *cross-examination*, asking questions about matters brought up during direct examination. The purpose of cross-examination is to test the recollection, knowledge, and credibility of a witness. In practice, however, it is applied to refute or at least cast doubt upon facts attested to during direct examination by catching the witness in a contradiction (*impeaching* the witness), by casting doubt on his or her character or capabilities, or by causing him or her to say or do something that causes mistrust in a jury.

An attorney will engage in *redirect examination* to correct misinterpretations of answers given during cross-examination or if the testimony of the witness has been impeached. When an expert witness is involved, impeachment will sometimes occur when the testimony at trial contradicts what was said at the deposition. In such instances opposing counsel usually will not give the expert an opportunity to explain why the difference exists. To *rehabilitate* his or her expert's testimony, the lawyer will ask for an explanation during redirect. If the plaintiff engages in redirect testimony, the defense can perform a *recross examination,* but the questions must be limited to the topics covered in the redirect examination.

7.9.1 Attempts to Disqualify the Expert Witness

There are at least three fronts on which the opposing counsel can attack an expert witness: credibility, integrity, and testimony. Often these three fronts are cited as credibility, credibility, and credibility.

The opposing counsel may try to convince the jury that the expert has given his or her opinion without adequate forethought or that the opinion was a judgment made in haste as well as instigated by the attorney. Another technique used by the opposing counsel is to get the expert to provide a *laundry list* of the facts in the case. Following the presentation of this list, the opposing counsel will tell the jury that the expert has left out an important reference; the implication is that the expert is unaware or unsure of important facts. Opposing counsel is, in fact, trying to get the expert flustered so that any subsequent answers to questions posed by opposing counsel are stammered, seemingly uncertain, and incomprehensible to the jury.

Opposing counsel may also try to make the expert give a single citation on which he or she relied to form the opinion. This is an attempt to prevent the jury from hearing that the expert conducted an extensive literature search before forming the opinion. An alternative ploy is for opposing counsel to ask the expert whether he or she is sure there are no more references related to the issue in question. The expert needs to be adamant that the references used are the ones relevant to the case and in forming his or her opinion.

On occasion, opposing counsel will bring a textbook to the trial and quote a passage from that book that apparently disagrees with any statements made by the expert witness. The expert needs to respond that textbooks are written to teach principles and the information in them is not applicable in all cases. Another ploy is for the opposing counsel to read a sentence or part of a sentence from the expert's deposition and inquire whether the expert still agrees with that statement. Opposing counsel may even suggest that the expert lied when answering a particular question at the deposition.

The expert should first ask to see the page or quote in the context in which it was written, read it carefully, and then respond accordingly. If the opposing counsel has suggested that the expert lied, it is permissible to state that the expert is under oath (showing the jury that he or she has not forgotten) and that the answers are the truth. Then the answer that the expert gave in the deposition should be reaffirmed. No elaboration is necessary unless the expert is asked to do so.

Opposing counsel can quote his or her own expert witness and say or imply that the attorney's expert is not as familiar with the facts as is the opposing counsel's expert. Another ploy is for the opposing counsel to listen to the expert's answer and then cut the expert short, thus permitting the jury to hear only part of the expert's testimony. In this situation, the expert can appeal to the judge to be permitted to finish the answer. At this point, the expert should make eye contact with various members of the jury.

Opposing counsel can question the validity of the results of any test or experiment conducted that is damaging to his or her case. Counsel can ask the expert to discuss the test in great detail. If the expert hesitates on any point, opposing counsel will use this opportunity to try to throw doubt on his or her scientific competence. In the best of all worlds, opposing counsel may realize that the expert does indeed know the issues involved in the test method protocols and thus avoid asking any questions of substance for fear of getting damaging answers.

7.9.2 Direct Examination

Before the trial, the expert should have acquainted the attorney with any publications the expert has written that can even remotely be related to the present case. With the sophisticated databases available, the opposing counsel can get a complete list of the expert's published work. There are also companies that can furnish photocopies of any or all of the expert publications.

During direct examination by the attorney, it is the expert's job to present evidence to the court in the form of clear and concise answers to the attorney's questions. This may leave the expert with a feeling of well-being. The expert should not be fooled by his or her performance under the umbrella of the friendly direct examination. No matter how well the witness feels about his or her direct examination, opposing counsel is now ready for the attack. He or she has sharpened his or her teeth and is about to cross-examine the expert.

Most likely, opposing counsel will have obtained copies of all depositions and court records in which the expert has testified. Of course, the expert should have reviewed with the attorney any statement made previously that could have some bearing on the case at hand. If there is some potentially embarrassing statement in the expert's earlier testimony, the expert and the attorney must decide how, within the boundaries of scientific integrity, the expert can explain it. If new information has appeared after the previous statements, conclusion, or opinions were recorded, the expert should be prepared to cite it.

The ideas that emerge from science and engineering are dynamic and not static. In contrast to what the expert testified in a previous case, new information may have come to light that required the expert to change his or her mind. This information could have come from new publications or even from unpublished data related by a scientific colleague. This possibility should be explained carefully, without giving the jury the feeling that the expert is defensive on this subject.

As mentioned previously, opposing counsel will undoubtedly ask the expert again and again how many times he or she has testified in court. At this point, opposing counsel may also ask what percentage of these cases was for the plaintiff and what percentage for the defense. Whatever the expert's answer, opposing counsel will try, by direct statement or innuendo, to convince the jury that the expert is not unbiased but philosophically favors one side over the other. Along this line of questioning, the opposing counsel may also ask about the number of times the expert has testified or evaluated cases for the expert's present attorney. He or she may also ask if the expert has done any work for another attorney in that attorney's law firm.

In all instances, the expert should not answer any question too quickly. Despite the potential for a negative impact on the basis that the expert may be not be sure of the answer, he or she is advised to pause before answering *any* question posed by opposition counsel, to give the expert's attorney time to voice any objection to the question. The expert should determine the attorney's preference beforehand so that the technique can be practiced during the testimony rehearsal. The expert should not be afraid to pause and should take whatever time is needed to think out his or her answer or overcome any emotional response. The expert should remember that the trial, like the deposition, is being recorded and any off-beat, hurried answers can serve opposing counsel well and influence the outcome of the dispute.

Above all, the expert should answer all questions (from his or her attorney and from opposing counsel) clearly, concisely, and without prejudice to show the jury that the expert does not have any bias but accepts cases on the basis of the technical issues under examination rather than on the basis of the personalities involved. By doing this, the expert will give a sense of honesty and reliability as the purveyor of impartial testimony to the judge and to the jurors.

7.9.3 Attitude and Demeanor

Attitude and demeanor are important aspects of the expert's performance on the witness stand. In fact, now that many depositions are being video recorded, the same applies to the expert in a deposition. There are many pointers for the scientist or engineer to follow to ensure that the correct attitude and demeanor and those that follow are applicable to both depositions and trials.

The expert should not be too *modest*. He or she has been retained because of recognized expertise in a particular branch of science or engineering. Many experts tend to downplay their accomplishments and, in so doing, will minimize credibility. By the same token, the expert should not be boastful and exaggerate his or her accomplishments. Exaggeration may appear to be lying, especially in a résumé; this can lead to the expert's exclusion from the case and even to a charge of perjury.

The expert should be conscious of his or her *body language*. The judge or jury will be looking at each witness intently, seeking signals that help indicate whether or not what is being said is said with or without conviction. If the expert has a physical disability from a past injury that causes him or her to move in the deposition chair or in the witness stand, the attorney should make sure that this is known to the judge and jury through a well-posed question prior to direct examination. Also, the expert should be aware of his or her *facial expressions,* which are a form of body language. For example, the expert should not scowl at opposing counsel. The expert should remember that the judge was a lawyer and may not take kindly to a negative attitude toward someone who merely is doing his or her job. It is not unknown for opposing counsel to have hired a profiler who will be present in the courtroom to expose the expert's weaknesses and then pass on these findings to the counsel to use during cross-examination.

The expert should *speak clearly*. Depending on the acoustics in the courtroom and whether or not a microphone is used, it may be necessary for the expert to speak more loudly than is customary. In all cases, the expert should be certain to speak clearly to ensure that what he or she says is properly recorded. The expert should remember that the case may be decided on the basis of the transcript, which must be an accurate recoding of what has been said during the trial.

At all times, the expert should *show respect to the judge*. An expert should not act in a flippant manner toward the judge. The judge should always be addressed as "Your Honor." Many scientists and engineers forget this and address the judge as "Sir," which is also acceptable, although less preferable. Showing respect to the judge also shows respect to the court and the proceedings.

The expert should also remain *dispassionate* and *cordial,* no matter what opposing counsel may say. The attack he or she may mount against the expert (the attorney is just doing his or her job) is nothing more than an act—often put on simply to evoke an emotional and uncontrolled response. No matter what is said, the expert should react to counsel's words in an objective, dispassionate, and cordial manner.

The expert should not attempt to engage in a *battle of wits* with the opposing counsel. The expert is present to render impartial testimony so that the judge and jurors can understand the technical issues. An expert who engages in a battle of wits with opposing counsel will appear to be an advocate and thereby damage his or her credibility. An expert who moves to engage in a battle of wits with the opposing counsel may only be playing into the hands of an effective cross-examiner.

The expert should *be responsive* insofar as he or she only answers the question that is asked. First, this prevents the expert from rambling on and volunteering other information and becoming a star witness for opposing counsel. Second, it will prevent the judge from having to issue the instruction to please answer the question, which can detract from the expert's professional image and credibility.

Experts sometimes wonder how they are doing or how the case is going and will adjust their testimony in an effort to improve the case from their perspective. The expert should not do, this because he or she may lose concentration. The expert should concentrate on each question asked. A distracted expert can give erroneous answers, and efforts made to correct the situation can destroy his or her credibility.

Finally, a number of do's and don'ts should be observed during the time the expert is seated in the courtroom:

Do not read a newspaper. Doing so shows obvious disrespect, and the judge may advise the expert to put it away. This is a discourtesy that will be remembered when the expert takes the stand. Pay attention to the proceedings and take notes.

Do not whisper to anyone. The judge may single out the expert to admonish for discourteous behavior.

Do not leave and enter the courtroom frequently. Wait for recesses to move about.

Do not give an opposing expert any effusive greeting, no matter how well he or she is known. This only enhances the opposing expert's image.

Do not engage in extensive note-passing to the attorney unless the matter is extremely urgent. Go over notes with the attorney during a recess, out of earshot of others.

Be cautious of what is said when not on the stand. In many instances members of the opposition team will be scattered around a courtroom, in hallways, and in corridors and may pick up an important piece of information dropped casually.

7.9.4 The Evidence of Others

There is a potential dilemma looming for the expert if opposing counsel asks if the expert knows anyone who agrees with his or her opinion. This type of question can be a real problem. There are competent, honest scientists who do not believe or agree with other competent scientists on generally accepted scientific inference. Because expert witnesses must interpret and have opinions on the facts of a case, it is not surprising that they disagree. Facts are often imprecise, and perceptions of facts often depend on mindset. Some scientists will state their disagreements openly. If the expert witness is asked to name those scientists who agree with his or her opinion or conclusion, there should be no problem naming other scientists whose views are in the public domain. Those agreeing scientists may have written about or given public speeches expressing their views.

If the expert does know firsthand of the unconventional views of another scientist, there is no choice, under oath, but to answer truthfully, even though the expert may create hard feelings later among the scientists named. Above all, it is not wise to make disparaging remarks about any other scientist or engineer. In the long run, when there is a battle of the expert witnesses, the jury declares the victor.

In many cases, to probe for weaknesses, opposing counsel may quote from a text with which the expert may not be familiar in order to ask the source of the quotation and also to indicate that the author of that particular text was not qualified to make such a statement. Another approach is for the expert to state that the work quoted is in complete disagreement with other authoritative work or texts in the field, remembering to cite the name and author of the text he or she is referencing.

In some cases, opposing counsel my use a book that is out of date. The expert should ask to see a copy of the book in question. If this is possible, he or she should check the front material and, if the book is out of date, the expert should make a statement for the record that the book is out of date and the field has advanced since the book was published. The expert should beware of the book that is an old edition with a new binder and bear in mind that even authoritative authors may have written something years ago that they have since contradicted. It is helpful for the expert to be familiar with various works to permit statements indicating the author has completely reversed from that early edition.

However, both the attorney and opposing counsel will have a problem with the expert witness who offers an unorthodox explanation for a scientific phenomenon. Many scientists and engineers may not follow mainstream thinking and can offer an alternative theory; other scientists may merely present off-the-wall hypotheses. Both the expert and the attorney must analyze with open minds any unorthodox statements and make major decisions about the validity of these explanations. The attorney can then handle such issues during direct examination.

If the expert has respect for an iconoclast who presents off-the-wall opinions, he or she should state this clearly; nonetheless, it is possible to disagree. The expert should state his or her opinion to the jury unemotionally, without rancor or even humor. At all times, the expert should be cautious and aware that the iconoclast may, in the long term, be right. At no time should any expert witness underestimate the opposing expert witness.

7.9.4.1 Hearsay Evidence

As noted elsewhere, subject to the judge's permission, it is may be permissible for an expert to give an opinion on *hearsay testimony,* which is normally excluded from court hearings. The hearsay testimony must be of a type usually relied upon by other experts in the field in which this expert is testifying. Thus, there is some leeway for the expert to express an opinion on the case. For example, unpublished information may be relied upon by the expert; however, it is advisable that an expert worth his or her salt not rely upon unpublished information to any great extent.

Many judges and juries are suspicious of hearsay evidence or unpublished data. The frequent use of such evidence (or data) can cast doubts upon what otherwise may be an excellent testimony. The difficulty is whether the court thinks the testifying expert has actually relied on the opinion of the other expert or whether the expert affirms that the other expert agrees with the testifying expert's opinion. A major problem occurs when the expert includes in his or her opinion information from a conversation with another expert. The testifying expert must demonstrate the validity of his or her analysis of the facts and the conclusions drawn from these facts.

7.9.4.2 Real Evidence

Judges and juries thrive on real evidence—that is, evidence that has substantial technical data to back it up and has been proven by several workers. In some cases, an expert may testify that the discussion or even the opinion he or she gives, although counter to mainstream science, is derived from substantiated and proven facts. The complication here is that some theories are sound, whereas others are *junk science* and are not based on any form of logic or reliable data.

It is often difficult for a lay group like a jury to distinguish between testimony involving junk science and the real facts behind the testimony. In addition, opposing counsel will usually have a witness to deny the claims and theories of the expert who uses junk science. When this happens, the judge and jury will, understandably, be suspicious of the expert; his or her remaining testimony, no matter how valid, is then subject to dismissal from the conscious thoughts of the jurors. The expert's credibility is disastrously undermined by use of noncredible scientific or engineering testimony, studies, or opinions. The expert will emerge from the witness stand having as much credibility as the carpetbagger of decades past clothed in a loudly colored plaid suit and white shoes.

7.9.5 Cross-Examination

A form of cross-examination occurs during a deposition, but the questions asked are posed basically to obtain information. However, cross-examination during a trial is performed not only to probe weak spots, but also to discredit an individual somehow. Thus, the objective of cross-examination is to bring to the jury the points that opposing counsel believes will strengthen his or her case and weaken the attorney's case. Often this means that the opposing counsel will try to nullify the testimony of the expert witness. If it is not possible to negate that testimony, the next best strategy is to weaken it. Opposing counsel may even try to lead the expert witness into a verbal labyrinth in an attempt to confuse the expert, ultimately leading to his or her disqualification. In many cases, the name of the game is not to present the truth to the judge and jury but to destroy the opposing expert witness and win the case.

The cross-examination is theoretically limited in scope to permit the opposing counsel and, through his or her questions, the jury to consider if the expert is qualified for the case. Thus, the very first thing the opposing counsel will certainly do is take advantage of any failure by the attorney to qualify the expert properly as an expert, especially in distinguishing the PhD or DSc from the MD—in spite of the fact that the opposing counsel already knows the expert's true educational background, which will have been discussed during the deposition. It is essential that this aspect of the expert's history be covered fully and correctly in the deposition.

During cross-examination, the expert should remain objective and not permit emotions to rise to the fore. It is important to be polite to the opposing counsel and refrain from using sarcasm or even humor. The expert should be aware that the opposing counsel may show or simulate hostility toward the expert. Answering in kind does not help the expert's position. A dignified response is impressive under all circumstances; above all, the expert should remain dignified before the jury. If the expert feels that self-control may be slipping, a request to the judge for a short break will almost always be granted.

The expert should be prepared to answer what may appear to be mundane or even very personal questions, in addition to inquiries such as the number of times the expert has testified in a trial and the number of times the expert has testified for the plaintiffs and the number of times for the defense. When answering such questions, the expert should consider describing the time span; approximate dates are acceptable. At any time, when appropriate, the expert may make it clear that he or she is a professional consultant, not a professional witness. Opposing counsel may often follow up with a question about what percentage of the expert's income is derived from testifying.

Opposing counsel or even the judge may ask for dates, especially the date when the expert was first approached by the attorney. The expert must be careful about the difference between first contact and when he or she was first asked to be a consultant, and then when he or she was designated to be the actual expert in the case.

Another important set of dates may be requested by the opposing counsel related to when the expert arrived at his or her opinion as well as a timetable of events and thoughts that led to the opinion. When specific dates are requested, it is permissible to refer to a set of notes or a date book. This is accepted practice; no one is expected to have all important dates in his or her head. It is wise to keep a log of events leading up to the expert's opinion.

During cross-examination, the expert should listen carefully to the opposing counsel; sometimes the strategy the opposing counsel is using to negate the expert's testimony will be apparent. If the opposing counsel poses a question on fundamental principles, the expert may agree, but he or she should be wary of a correlation that is not germane to the case under trial. It is important to disagree without getting into an argument so that the opposing counsel cannot tell the jury that the expert is an advocate and not an impersonal, independent witness. Some questions may require a limited but not complete answer, and some questions may have many clauses, qualifying adjectives, and adverbs. Do not hesitate to ask opposing counsel to clarify the question. Any problems with awkward questions can be clarified on redirect examination by an alert attorney.

The expert should be wary of double-negative questions. Another common tactic is for the opposing counsel to pause and gaze at the witness after a brief reply to a question; many experts feel pressure to say something in addition. The expert should not succumb, for this can lead to problems.

The expert should avoid volunteering information beyond what answers are needed for the questions asked by either the attorney or opposing counsel. If by chance the expert mentions an extraneous topic or if the experts elaborate too much and gives too many examples in answer to a direct question, the opposing counsel can bring these topics up in cross-examination. Diversions can be the subject of cross-examination. If the expert has ever written a paper for scientists or for the lay public, the opposing counsel is free to ask about it without showing that paper to the expert. There have been times when an opposing counsel asked an expert to comment on a scientific journal article not written by the expert. The expert has every right to request a copy of that paper and can take as much time as necessary to read and think about it.

One last major point to ponder is that many opposing counsels have cross-examined experts in such a way as to enhance the opposing counsel's case. This is because the opposing counsels are successful in confusing the expert—twisting the interpretation of what the expert says, or just outwitting the expert.

In all probability, the opposing counsel will have a laptop computer available. As noted, all testimony is taken verbatim, and the transcript of the court proceedings can be available in real time; therefore, it may be possible for the opposing counsel to scan back to a previous answer to a question and get the exact wording of the answer. This information will be used to try to embarrass the expert on any discrepancies. Any answer given during cross-examination that is not precisely the same as that given before will be challenged.

The opposing counsel may also try to impeach the expert's testimony by quoting from his or her testimony in a different lawsuit. Even if the attorney did not obtain the transcript from the expert during the expert's deposition, he or she may have received it from other lawyers or organizations.

In summary, the witness should always answer questions posed by the opposing counsel in a calm manner and not give the appearance of being defensive. Objectivity and credibility are the goals to be achieved, even when opposing counsel uses tactics to get the witness. Some attorneys will object to the court on the grounds that opposing counsel is *badgering* or *harassing* the witness; even though this objection is often overruled, it gives the witness time to regain his or her composure.

If the judge asks any questions, the witness should turn in the chair and look directly at the judge while answering. If the trial is in the district court before a jury, the witness should look at the jury as much as possible while testifying. If the witness is nervous and cannot think of anything else, he or she may merely look at the person posing the questions. However, if a witness has prepared a good report, has thorough knowledge of the subject and issues, and is well dressed and rested, he or she is more likely to be calm, comfortable, and confident. Preparation such as this will present the witness with the best chance of success.

7.9.6 The Abusive Attorney and Suitable Responses

As mentioned, during the cross-examination, the opposing counsel will attempt in any way possible to nullify the information the expert gave during both the deposition and the direct examination in court. The main objective of the opposing counsel is to display the expert (and hence the expert's opinion) to the jury in the most unfavorable light possible. Misleading questions may be used, and sometimes the opposing counsel will say or insinuate that the expert came to his or her opinion hastily, without considering all facts and factors of the case.

In some cases, the opposing counsel will be abusive. The expert must, at all costs, remain calm. If the expert is polite at all times, this action can sometimes defuse a vicious attack or cause it to backfire on the hostile opposing counsel.

One hint: the expert should avoid looking at his or her attorney when answering a question posed by the opposing counsel. The jury or judge will think the expert is looking to the attorney to give the expert a clue or to see if the expert's attorney accepts or rejects the expert's answer. Also, the expert should avoid humor and sarcasm.

Occasionally, the following situation has occurred: the opposing counsel not only deals with an expert in a hostile manner but also becomes verbally abusive by inquiring about the expert's personal life, especially finances, or by alluding to sexual behavior. More often than not, the abusive opposing counsel will try in any way possible to make the expert uncomfortable. This can be done, especially during the deposition, by making the room uncomfortable, delaying needed breaks, or otherwise trying to manipulate the proceedings so that the expert will suffer real fatigue.

Opposing counsels can ask insignificant questions, can be repetitive and argumentative, and may ask personal questions not related to the case. They can attempt to force a witness to render an opinion or elaborate on an opinion beyond the actual situation or facts. An expert witness has a right not be abused at the deposition or trial. The expert should be definite and open in stating, in a polite manner, that the question asked is offensive. Above all, the expert can refuse to answer an abusive question; this must be done, if possible, without engaging in a mud-slinging contest. The expert who gets into a fight with the opposing counsel usually loses face with the jury; the jury quite often may ignore this expert's testimony. All this must be on the record.

7.9.7 Redirect Examination

At the conclusion of cross-examination, the attorney who called the witness will be given the opportunity for redirect questioning. If the attorney believes the opposing counsel has brought out certain points (or facts) that will either mislead the jury or make the expert's original statements less clear, the attorney may wish to ask the expert further questions on redirect examination.

Redirect questions should be designed to allow the witness to amplify or expand upon subjects raised during the cross-examination. Redirect or recross examination must relate to questions asked or the subjects raised by opposing counsel. Often, the parties will stipulate that each other's witnesses are experts and the reports are also stipulated as the experts' reports. In such a case, the opposing counsel will proceed to the cross-examination point as soon as the expert is put under oath on the witness stand.

Generally, new facts are not brought out at this time. The attorney may just wish to emphasize points made earlier in the direct examination. This is also the opportunity for the attorney to counteract any wrong impression made to the jury during the cross-examination. In other words, the redirect is to clear up various points that the attorney believes were left unclear.

7.9.8 Recross Examination

The opposing counsel has one more chance to diminish the effect of the expert's testimony on the jury. As a rule, the recross examination must be confined to those aspects covered in redirect examination. New evidence and new types of questions are not permitted. It must again be emphasized that the expert witness must not elaborate too much when answering a question because the expert may give the opposing counsel an opening to introduce other questions. If the attorney objects, opposing counsel can prevail on the basis that it is not a new topic because it was introduced by the witness.

8 Epilogue

8.1 INTRODUCTION

An expert's obligation does not stop when the verdict is given by the judge or jury, or even if the judge dismisses the suit (summary judgment). There are many details yet to be completed.

A scientist's or engineer's future as an expert witness depends on a number of factors. The relationship with the attorney is important, and this depends upon the expert's performance during the preparation of the case and at the trial. Finally, much depends now on the expert's willingness to be subjected to the tribulations of a trial again, to an opposing counsel who may be abusive, and to a judge who may be unsympathetic.

The expert should be concerned enough about his client to call the attorney after the trial is over to learn the final verdict. Although this may seem to be common curiosity, the conversation creates an opportunity for the expert to request that the attorney give a commentary on the expert's performance.

The matter of disposition of materials also is a concern. If the expert is, for any reason, requested to retain certain items, he or she should do so. On the other hand, if the expert is given the authority to discard materials, it is suggested that he or she write a letter to the attorney confirming these instructions.

However a case is not necessarily closed after final judgment is entered. Either party (usually the party who lost) can enter an appeal to an *appellate court* because errors were made at trial, because the award was excessive, or for other reasons. The *appellant* must notify the clerk of the court where the original trial was held that an appeal is being filed and must notify the *appellee* as well. Sometimes, an appeal is filed principally to delay payment, in that it bars the appellee from enforcing judgment until the appeal is heard and a verdict is rendered. Accordingly, the appellant must post a bond to cover costs and the final judgment, in case the appeal fails and the appellant's assets are dissipated in the interval.

The appellant's next step is to prepare a *record of appeal,* which identifies exactly why the appeal should be granted, citing whatever precedents may apply. The appellee also files a record of appeal, usually to prove the original verdict should stand.

The appellate court consists of several judges, and the court's decision is based on their majority opinion. They may affirm the lower court's ruling, reverse or modify the judgment, or grant a new trial. Once the judges reach their decision, the case is sent back to the trial court for whatever action is necessary.

In some rare instances, the verdict of an appeals court can be appealed, but only when a particularly difficult or contentious question of law exists. In states that have an intermediate appellate court, an appeal from its decision would be taken to the highest court in the state, usually called the state supreme court or supreme appellate court. Even that decision can be appealed by seeking redress in a federal district court or even with the Supreme Court of the United States.

8.2 AFTER-TRIAL RESPONSIBILITIES

An expert must see to it that myriad details are taken care of and all tasks satisfactorily completed. It is essential to contact the attorney after the trial or after testifying. If possible, the expert and the attorney should have a postmortem session in person. If this is not possible in a reasonable time, a courtesy letter is in order. The expert should thank the attorney (and vice versa) and, if there have been any problems in the relationship, include a diplomatic note suggesting possible remedies. Perhaps the expert can make a checklist of good and bad points and share it with the attorney. At this time, it is wise to include any additional bills and inquire when payment can be expected.

Too often, the attorney does not contact the expert immediately at a trial's end to let the expert know that a verdict was given. It is nice to learn whether the attorney believes that the expert's testimony had significant impact on the favorable outcome of a case.

If a particular case or trial appears to be unusual to the expert, or even to the attorney, the expert may consider writing an academic article describing the distinctive aspect of the case. The attorney must make the final decision as to whether an article is in order. No writing should be done until the final disposition of the case, after all appeals are completed. If the expert does not heed this warning and publishes a paper and the case is retried for any reason, this published article will place the expert in an unfavorable position.

No matter on which side the expert has testified and offered an opinion, if there is publicity about the trial, some scientific colleague or other will be critical of the expert's testimony. Often, the critics will get the facts wrong from the mass media or will somehow have an incorrect perception of the testimony. Not all scientists will take the time to evaluate the validity of the expert's testimony. A small minority will not criticize the science proposed but, rather, the scientist for testifying.

8.3 THE TRIAL TRANSCRIPTS

There is no agreement on how long to keep papers and notes pertaining to the trial. Some experts keep everything forever. Others clear their files as soon as they know of the disposition of a case. One expert has a five-year rule: if there has been no activity on a particular case for five years, then all materials pertaining to it may be destroyed. It is a good general rule to keep all records at least until after all appeals have been resolved. Often, a year may not be sufficient time for a possible new trial; thus, all papers pertaining to the case may have to be kept for two years or longer. Some attorneys ask experts to return the papers. If this is the case, return all materials by certified mail, with a return receipt requested. The receipt is the expert's protection.

The expert must contact the attorney in regard to the disposal of any papers or property. If the attorney requests that the expert have all documents destroyed, the destruction of all papers must be in such a manner that they cannot be retrieved by anyone. A shredder is best, although personally taking the papers to the dump is acceptable. If possible, papers should be burned.

If the attorney leaves the issue of the trial-related documents to the expert, the expert must make a decision whether to keep them all or only part of the collection. It is possible that these papers can be obtained by a future opposing counsel through discovery in a different case. It is a good idea, however, to keep copies of all the expert's personal depositions. Also, the expert should keep a file on all the trials in which he or she was involved—their dates, issues of the trial, the sides for which the expert testified, and the names of the attorneys and firms.

8.4 FUTURE LITIGATION ARISING FROM THE TRIAL

As enunciated in Chapter 1, a controversy before a court or a lawsuit is commonly referred to as litigation. If it is not settled by agreement between the parties, it will eventually be heard and decided by a judge or jury in a court. However, the case may not stop at the stage where the jurors have heard the evidence, the jury passed down a verdict against the defendant, and the judge pronounced sentence. Being extremely dissatisfied with the judge's ruling, the defendant may decide to appeal.

An *appeal* is a formal request that a higher body—typically a higher court—review the action, procedure, or decision of a lower court, administrative agency, or other legal or judicial body. An appeal is normally made by the party who loses or did not get all the relief that was sought. If both parties are dissatisfied, each may appeal part of the decision.

Trials at which witnesses and other evidence are presented to a jury or judge in order to determine the truth or facts regarding a particular case are held only in courts with original jurisdiction (i.e., courts in which a lawsuit is originally and properly filed, which have the power to accept evidence from witnesses and make factual and legal determinations regarding the evidence presented). These trial courts also determine what, if any, punishment (in criminal cases) or damages (in civil cases) should be meted out.

On the other hand, *courts of appeal* (*appeals courts*) possess only appellate jurisdiction; they do not hold trials. Instead, these courts review decisions of trial courts for errors of law. Accordingly, an appeals court considers only the record (i.e., the papers the parties filed and the transcripts and any exhibits from any trial) from the trial court and the legal arguments of the parties. These arguments, which are presented in written form, can range in length from dozens to hundreds of pages (*briefs*). Sometimes, lawyers are permitted to supplement the written briefs with oral arguments before the appeals judges. However, at such hearings, only the lawyers are allowed to speak to the court. The expert may be present to offer advice to the attorney.

The Federal Rules of Appellate Procedure govern the procedure in the courts of appeals. In a court of appeals, an appeal is almost always heard (for the current context) by a panel of three judges who are randomly selected from the available judges (including senior judges and judges temporarily assigned to the circuit).

Historically, certain classes of cases held a right of automatic appeal to the Supreme Court of the United States. For example, one of the parties to the case could appeal a decision of a court of appeals and take the case to the Supreme Court. There is no longer any right of automatic appeal for a decision of a court of appeals, but a party may apply to that court to review a ruling of the circuit court—called petitioning for a writ of certiorari—and the Supreme Court may, in its own discretion, review any such ruling.

Appeals court decisions, unlike trial court decisions, are binding precedent. Other courts in that circuit must, from that point forward, follow the appellate court's guidance in similar cases, regardless of whether the trial judge believes that the case should be decided differently. Laws may change over time; therefore, the law that exists at the time of the appeal may differ from the law that existed at the time of the events being disputed by the litigants. A court of appeals applies the law as it exists at the time of the appeal; otherwise, it would be handing down decisions that were instantly obsolete.

8.5 RECORDS ACCUMULATED DURING THE CASE

It is a personal reference, but many experts prefer to retain copies of all their personal depositions as well as any related abstracts of trial transcripts. They also maintain a file of the trials in which they were involved as well as the names of the attorneys, opposing counsels, and respective law firms. However, the main issue arises when an expert looks at the boxes of other records that he or she has accumulated during the trial. However interesting they may be in the initial days or weeks after the trial, there is always the question of retention or disposal of these records.

There is no agreement on how long to keep scientific documents, legal documents, electronic records, and other papers and notes pertaining to the trial. Some experts keep everything within a time limit (say, five years), while others do not have a time limit. Other expert witnesses clear their files as soon as they are informed of the judge's ruling, usually because of limited storage space.

Immediate destruction of case-related files after the judge's ruling can be convenient, in terms of saving space, but can also be costly. Many cases tried in a court of law are subject to appeal, in which case the expert needs files once more to refresh his or her memory on the salient features of the case. Many experts feel that if there has been no action on a case for one year and all of the appeals have been resolved, it is then time to dispose of the case-related records. Because most cases require more than a year, the expert is advised to be cautious and keep all papers and records pertaining to the case for two years or even longer.

If there has been no activity on a particular case for five years, all materials pertaining to it may be destroyed. Nevertheless, related cases can come to light just within or shortly after a five-year period has lapsed and the same attorneys prefer to retain (subject to prior satisfactory performance) the same experts.

However, many attorneys will ask the expert to destroy all paper and electronic files at the termination of the case (i.e., when the judge's decision is known). On the other hand, some attorneys ask the experts to return all of the papers and electronic files. If the expert receives this request, he or she should return all materials by certified mail, with return receipt requested.

If the attorney requests that the expert have all documents destroyed, the destruction of all papers must be in such a manner that they cannot be retrieved by anyone. Paranoia aside, a cross-cut shredder is best because the papers cannot be reconstructed. Many shredders also have the ability to shred CD-ROM and DVD disks. Personally transporting the papers to the local recycling center is not always acceptable, unless the expert can remain on site to see that papers are destroyed in the recycling unit. Transportation to the local landfill site may also be acceptable. The best of all worlds is a trip to the local incinerator, where the expert (hopefully) can remain on site to observe the papers and disks being burned. The worst of all worlds is when the attorney is nonchalant and allows the expert to dispose of the records or retain them according to personal choices.

Scientists and engineers will usually wish to retain scientific and engineering papers for their own files and libraries. The nontechnical documents should be disposed of by any of the preferable methods given earlier. However, it is possible that all or any of the documents can be obtained by a future attorney or opposing counsel through discovery in a related or even different case.

8.6 LESSONS LEARNED

It is at this time that the seasoned (i.e., multiple cases) or recent (i.e., first case) expert should sit back and muse over the lessons learned from the recent trial (and possibly the appeal). Several questions and corresponding answers may come to mind, some of which (in a light-headed moment) might be:

Thoughts while watching other experts under cross-examination: I want to go home

Most important thought during cross-examination: I want to go home

Most memorable words of the trial: "No further questions, your Honor," as opposing counsel closes cross-examination

Most memorable moment of the trial: dismissal by the judge after giving successful testimony

Consistent thought during trial: hoping the judge and jurors understand science and engineering

Method used to relax at night during the period of the trial: watching Clint Eastwood in the movie *Hang 'em High*

Books read during the day while waiting to testify: Great Expectations and *A Tale of Two Cities,* by Charles Dickens

One thing that may not surprise people about the expert: instead of a scientist or engineer, he or she wanted to be liberal arts major
Reason why the expert has remained in science or engineering: nowhere else to go
Biggest challenge for the future: deciding whether or not to take on another case
Highlight of the trial: learning that a decision has been made in favor of the client and attorney who retained the expert
Celebration of a win: thinking "Is that all there is?"
Alternate method of dealing with a win: start breathing normally
Dealing with a loss: thinking "Is that all there is?"
Alternate method of dealing with a loss: start breathing normally

Glossary

This glossary is by no means comprehensive; it is a compendium of words that the scientist or engineer is likely to come across during encounters with attorneys and the judicial system. For further enlightenment, the glossary also contains a selection of Latin words and phrases used frequently in legal language and judicial proceedings.

Abatement of action: A suit that has been quashed and ended.

Abstract: A summary of a larger work wherein the principal ideas of the larger work are contained.

Abstract of record: A short, abbreviated form of the case as found in the record.

Abstract of title: A chronological summary of all official records and recorded documents affecting the title to a parcel of real property.

Accessory: A person who aids in or contributes to the commission of a crime.

Accomplice: One who knowingly, voluntarily, and intentionally unites with the principal offender in the commission of a crime; a partner in a crime.

Accord: A satisfaction agreed upon between the parties in a lawsuit that bars subsequent actions on the claim.

Accord and satisfaction: A method of discharging a claim upon agreement by the parties to give and accept something in settlement of the claim.

Accusation: A formal charge against a person to the effect that he has engaged in a punishable offense.

Accused: The generic name for the defendant in a criminal case.

Acknowledgment: 1. A statement of acceptance of responsibility. 2. The short declaration at the end of a legal paper showing that the paper was duly executed and acknowledged.

Acquit: 1. To certify legally the innocence of one charged with a crime. 2. To set free, release, or discharge from an obligation, burden, or accusation. 3. To find a defendant not guilty in a criminal trial.

Acquittal: 1. In criminal law, a finding of not guilty. 2. In contract law, a release, absolution, or discharge from an obligation, liability, or engagement.

Action: Case, cause, suit, or controversy disputed or contested before a court of justice.

Action in personam: An action against the person, founded on a personal liability. In contrast to action *in rem* (q.v.), an action for the recovery of a specific object, usually an item of personal property such as an automobile.

Action in rem: Proceeding *against the thing* as compared to personal actions (*in personam*, q.v.); usually a proceeding where property is involved.

Additur: An increase by a judge in the amount of damages awarded by a jury.

Adjudicate: To determine finally.

Adjudication: Giving or pronouncing a judgment or decree; also, the judgment given.

Ad litem: A Latin term meaning for the purposes of the lawsuit. For example, a guardian *ad litem* is a person appointed by the court to protect the interests of a minor or legally incompetent person in a lawsuit.

Administrator: 1. One who administers the estate of a person who dies without a will. 2. A court official.

Admissible: Pertinent and proper to be considered in reaching a decision.

Admissible evidence: Evidence that can be legally and properly introduced in a civil or criminal trial.

Admission: Voluntary acknowledgment of the existence of certain facts relevant to the adversary's case.

Admonish: To advise or caution. For example, the court may caution or admonish counsel for wrong practices.

Adoption: To take into one's family the child of another and give him or her the rights, privileges, and duties of a child and heir.

Adversary system: The trial method used in the United States and some other countries. This system is based on the belief that truth can best be determined by giving opposing parties full opportunity to present and establish their evidence and to test by cross-examination the evidence presented by their adversaries. All this is done under the established rules of procedure before an impartial judge or jury.

Affiant: A person who makes and signs an affidavit.

Affidavit: A written statement of facts confirmed by the oath of the party making it, before a notary or officer having authority to administer oaths. For example, in criminal cases, affidavits are often used by police officers seeking to convince courts to grant a warrant to make an arrest or a search. In civil cases, affidavits of witnesses are often used to support motions for summary judgment.

Affirmative defense: Defense in which, without denying the charge, the defendant raises circumstances such as insanity, self-defense, or entrapment to avoid civil or criminal responsibility.

Affirmed: In the practice of appellate courts, the word means that the decision of the trial court is correct.

Agent: One who has authority to act for another.

Agreed statement of facts: A statement of all important facts, which all the parties agree is true and correct, that is submitted to a court for ruling.

Agreement: 1. A mutual understanding and intention between two or more parties. 2. The writing or instrument that is evidence of an agreement. (Although often used as synonymous with *contract*, *agreement* is a broader term.)

Alford plea: A special type of guilty plea by which a defendant does not admit guilt but concedes that the state has sufficient evidence to convict; normally made to avoid the threat of greater punishment.

Alibi: A defense claim that the accused was somewhere else at the time a crime was committed.

Alimony: A court-ordered allowance that one spouse pays the other spouse for maintenance and support while they are either separated, pending suit for divorce, or divorced.

Allegation: The assertion of a party to an action, setting out what he expects to prove.

Allege: To state, recite, assert, claim, maintain, charge, or set forth. To make an allegation.

Alleged: 1. Asserted to be true as depicted. 2. A person who is accused but has not yet been tried in court.

Allocution: A defendant's statement in mitigation of punishment.

Alternative dispute resolution (ADR): Settling a dispute without a full, formal trial. Methods include mediation, conciliation, arbitration, and settlement, among others.

Amend: To change, correct, revise, improve, modify, or alter.

Amendment: The correction of an error admitted in any process.

Amicus curiae: A friend of the court. One, not a party to a case, who volunteers to offer information on a point of law or some other aspect of the case to assist the court in deciding a matter before it.

Annotation: A case summary or commentary on the law cases, statutes, and rules illustrating its interpretation.

Annual review: Yearly judicial review, usually in juvenile dependency cases, to determine whether the child requires continued court supervision or placement.

Answer: The defendant's response to the plaintiff's allegations as stated in a complaint. An item-by-item, paragraph-by-paragraph response to points made in a complaint; part of the pleadings.

Appeal: A request made after a trial, asking another court (usually the court of appeals) to decide whether the trial was conducted properly. To make such a request is *to appeal* or *to take an appeal.*

Appearance: A coming into court as party or interested person or as a lawyer on behalf of party or interested person.

Appellant: The party appealing a final decision or judgment.

Appellate court: A court having jurisdiction to hear appeals and review a trial court's procedure.

Appellate jurisdiction: The appellate court has the right to review and revise the lower court decision.

Appellee: The party against whom an appeal is taken. Sometimes called a respondent.

Arbitration: The referral of a dispute to an impartial third person chosen by the parties to the dispute, who agree in advance to abide by the arbitrator's award issued after a hearing at which both parties have an opportunity to be heard.

Argument: Remarks addressed by an attorney to a judge or jury on the merits of a case or on points of law.

Arraign: The procedure where the accused is brought before the court to hear the criminal charge or charges against him or her and to enter a plea of guilty, not guilty, or no contest.

Arraignment: A proceeding in which the accused is brought before the court to plead to the criminal charge in the indictment or information. The charge is read to him or her and he or she is asked to plead guilty or not guilty or, where permitted, *nolo contendere* (no contest) (q.v.); another term for preliminary hearing.

Arrest: To deprive a person of his liberty by legal authority.

Arrest of judgment: Postponing the effect of a judgment already entered.

Arson: The malicious burning of someone else's or one's own dwelling or of anyone's commercial or industrial property.

Assault, aggravated: An assault committed with the intention of committing some additional crime.

Assignee: The person to whom property rights or power is transferred by another, a grantee.

Assumption of risk: In tort law, a defense to a personal injury suit. The essence of an affirmative defense is that the plaintiff assumed the known risk of whatever dangerous condition caused the injury.

At issue: The time in a lawsuit when the complaining party has stated his or her claim, the other side has responded with a denial, and the matter is ready to be tried.

Attachment: Taking a person's property to satisfy a court-ordered debt.

Attempt: An endeavor or effort to perform an act or accomplish a crime that is carried beyond preparation but lacks execution.

Attest: To bear witness to; to affirm to be true or genuine; to certify.

Attorney: Attorney-at-law, lawyer, counselor-at-law; in the current context, the lawyer who has hired or retained the expert witness to provide information or give testimony in the case on behalf of the *plaintiff* or the *defendant.*

Attorney-at-law: An advocate, counsel, or official agent employed in preparing, managing, and trying cases in the courts.

Attorney–client privilege: The privilege of a client to have his or her communications with his or her attorney protected. Based upon this privilege, an attorney must protect confidential information about his or her client or withhold information the client has given in confidence.

Attorney-in-fact: A private person (who is not necessarily a lawyer) authorized by another to act in his or her place, either for some particular purpose, such as to do a specific act, or for the transaction of business in general, not of legal character. This authority is conferred by an instrument in writing, called a letter of attorney or, more commonly, a power of attorney.

Attorney of record: The lawyer who represents a client and is entitled to receive all formal documents from the court or from other parties. Also known as counsel of record.

Authenticate: To give authority or legal authenticity to a statute, record, or other written instrument.

Bailiff: A court officer who has charge of a court session in the matter of keeping order and has custody of the jury.

Bankrupt: The state or condition of a person who is unable to pay his or her debts as they are or become due.

Bankruptcy: Refers to statutes and judicial proceedings involving persons or businesses that cannot pay their debts and seek the assistance of the court in getting a fresh start. Under the protection of the bankruptcy court, debtors may be released from or *discharged* from their debts, perhaps by paying a portion of each debt. Bankruptcy judges preside over these proceedings. The person or business with the debts is called the debtor, and the people or companies to whom the debtor owes money to are called creditors.

Bar: 1. Historically, the partition separating the general public from the space occupied by the judges, lawyers, and other participants in a trial. 2. More commonly, the term means the whole body of lawyers.

Bar examination: A state examination taken by prospective lawyers in order to be admitted and licensed to practice law.

Bench: The seat occupied by judges in courts.

Bench conference: A meeting at the judge's bench, either on or off the record, of the judge, counsel, and sometimes the defendant, out of the hearing of the jury.

Bench trial: Trial without a jury in which a judge decides the facts.

Bench warrant: An order issued by a judge for the arrest of a person.

Beneficiary: Someone named to receive property or benefits in a will. In a trust, a person who is to receive benefits from the trust.

Bequeath: To give a gift to someone through a will.

Bequests: Gifts made in a will.

Best evidence: Primary evidence; the best evidence available. Evidence short of this is *secondary*. For example, an original letter is *best evidence*, and a photocopy is *secondary evidence*.

Beyond a reasonable doubt: The standard in a criminal case requiring that the jury be satisfied to a moral certainty that every element of a crime has been proven by the prosecution. This standard of proof does not require that the state establish absolute certainty by eliminating all doubt, but it does require that the evidence be so conclusive that all reasonable doubts are removed from the mind of the ordinary person.

Bias: Inclination, bent, a preconceived opinion or a predisposition to decide a cause or an issue a certain way.

Bill of particulars: A statement of the details of the charge made against the defendant.

Breach: The breaking or violating of a law, right, obligation, or duty by either doing an act or failing to do an act.

Breathalyzer test: Test to determine content of alcohol in one arrested for operating a motor vehicle while under the influence of alcohol by analyzing a breath sample.

Bribe: A gift, not necessarily of monetary value, given to influence the conduct of the receiver.

Brief: A written statement prepared by the counsel arguing a case in court. It contains a summary of the facts of a case, the pertinent laws, and an argument of how the law applies to the facts supporting counsel's position.

Burden of proof: The obligation of a party to establish by evidence a requisite degree of belief concerning a fact in the mind of the trier of fact or the court.

Calendar: List of cases scheduled for hearing in court.

Calling the docket: The public calling of the docket or list of causes at commencement of term of court, for setting a time for trial or entering orders.

Caption: The heading on a legal document listing the parties, the court, the case number, and related information.

Case: A general term for an action, cause, suit, or controversy brought before the court for resolution.

Case law: Law established by previous decisions of appellate courts, particularly the Supreme Court.

Case number: See *docket number.*

Causation: The act that produces an effect.

Cause: A lawsuit, litigation, or action. Any question, civil or criminal, litigated or contested before a court of justice.

Cause of action: The facts that give rise to a lawsuit or a legal claim.

Caveat: A warning; a note of caution.

Caveat emptor: *Let the buyer beware.* Encourages a purchaser to examine, judge, and test for himself.

Cease and desist order: An order of an administrative agency or court prohibiting a person or business from continuing a particular course of conduct.

Certification: 1. Written attestation. 2. Authorized declaration verifying that an instrument is a true and correct copy of the original.

Certified: Attested as being true or an exact reproduction.

Certiorari: A means of getting an appellate court to review a lower court's decision. The loser of a case will often ask the appellate court to issue a *writ of certiorari* (q.v.), which orders the lower court to convey the record of the case to the appellate court and to certify it as accurate and complete. If an appellate court grants a *writ of certiorari*, it agrees to take the appeal. This is often referred to as granting cert.

Chain of custody: An accounting for the whereabouts of the tangible evidence from the moment it is received in custody until it is offered as evidence in court.

Challenge: An objection, such as when an attorney objects at a hearing to the seating of a particular person on a civil or criminal jury.

Challenge for cause: Objection to the seating of a particular juror for a stated reason (usually bias or prejudice for or against one of the parties in the lawsuit). The judge has the discretion to deny the challenge. This differs from peremptory challenge.

Chambers: A judge's private office. A hearing in chambers takes place in the judge's office outside the presence of the jury and the public.

Change of venue: Moving a lawsuit or criminal trial to another place for trial.

Character evidence: The testimony of witnesses who know the general character and reputation of a person in the community in which he or she lives. It may be considered by the jury in a dual respect: (1) as substantive evidence upon the theory that a person of good character and reputation is less likely to commit a crime than one who does not have a good character and reputation, and (2) as corroborative evidence in support of a witness's testimony as bearing upon credibility.

Charge: A formal allegation, as a preliminary step in prosecution, that a person has committed a specific offense, which is recorded in a complaint, information, or indictment; to charge; to accuse. See *instructions.*

Charge to the jury: The judge's instructions to the jury concerning the law that applies to the facts of the case on trial.

Charging document: A written accusation alleging a defendant has committed an offense. Includes a citation, an indictment, information, and a statement of charges.

Chief judge: Presiding or administrative judge in a court.

Citation: A reference to a source of legal (or scientific or engineering) authority.

Civil action: Noncriminal case in which one private individual or business sues another to protect, enforce, or redress private or civil rights.

Civil case: A lawsuit brought to enforce, redress, or protect private rights or to gain payment for a wrong done to a person or party by another person or party. In general, all types of actions other than criminal proceedings.

Civil procedure: The rules and process by which a civil case is tried and appealed, including the preparations for trial, the rules of evidence and trial conduct, and the procedure for pursuing appeals.

Claim: The assertion of a right to money or property.

Class action: A lawsuit brought by one or more persons on behalf of a larger group.

Clear and convincing evidence: Standard of proof commonly used in civil lawsuits and in regulatory agency cases. It governs the amount of proof that must be offered in order for the plaintiff to win the case.

Clerk: Officer of the court who files pleadings, motions, judgments, etc.; issues process; and keeps records of court proceedings.

Closing argument: The closing statement, by counsel, to the trier of facts after all parties have concluded their presentation of evidence.

Code: A collection, compendium, or revision of laws, rules, and regulations enacted by legislative authority.

Code of Federal Regulations: The CFR is the annual listing of executive agency regulations published in the daily *Federal Register* and the regulations issued previously that are still in effect. The CFR contains regulatory laws governing practice and procedure before federal administrative agencies.

Code of Professional Responsibility: The rules of conduct that govern the legal profession; general ethical guidelines and specific rules written by the American Bar Association.

Codicil: An amendment to a will.

Collateral: 1. Property that is pledged as security against a debt. 2. A person belonging to the same ancestral stock (a relation) but not in a direct line of descent.

Collateral attack: An attack on a judgment other than a direct appeal to a higher court.

Commissioner: A person who directs a commission; a member of a commission. The officer in charge of a department or bureau of a public service.

Common law: The legal system that originated in England and is now in use in the United States; based on judicial decisions rather than legislative action.

Complainant: The party who complains or sues; one who applies to the court for legal redress. Also called the plaintiff.

Complaint: 1. The legal document that usually begins a civil lawsuit. It states the facts and identifies the action the court is asked to take. 2. Formal written charge that a person has committed a criminal offense.

Comply: To act in accordance with, to accept, to obey.

Conciliation: A form of alternative dispute resolution in which the parties bring their dispute to a neutral third party, who helps lower tensions, improve communications, and explore possible solutions. Conciliation is similar to mediation, but it may be less formal.

Concurrent jurisdiction: The jurisdiction of two or more courts, each authorized to deal with the same subject matter.

Confession: Voluntary statement made by one who is a defendant in a criminal trial, which, if true, discloses his or her guilt.

Confiscate: To seize or take private property for public use. ("The police confiscated the weapon.")

Conflict of interest: 1. A real or seeming incompatibility between one's private interests and one's public or fiduciary duties. 2. A real or seeming incompatibility between the interests of two of a lawyer's clients, such that the lawyer is disqualified from representing both clients if the dual representation adversely affects either client or if the clients do not consent.

Consideration: The cause, price, or impelling influence that induces a party to enter into a contract.

Contempt of court: The finding of the court that an act was committed with the intent of embarrassing the court, disobeying its lawful orders, or obstructing the administration of justice in some way.

Continuance: The adjournment or postponement of a session, hearing, trial, or other proceeding until a future date.

Contract: A legally enforceable agreement between two or more competent parties made either orally or in writing.

Controlled substance: Any of the drugs whose production and use are regulated by law, including narcotics, stimulants, and hallucinogens.

Conviction: A judgment of guilty following a verdict or finding of guilty, a plea of guilty, or a plea of *nolo contendere*.

Corpus delecti: Body of the crime. The objective proof that a crime has been committed. It sometimes refers to the body of the victim of a homicide or to the charred shell of a burned house, but the term has a broader meaning. For the state to introduce a confession or to convict the accused, it must prove a *corpus delicti*—that is, the occurrence of a specific injury or loss and a criminal act as the source of that particular injury or loss.

Corroborate: To support with evidence or authority; make more certain.

Corroborating evidence: Supplementary evidence that tends to strengthen or confirm the initial evidence.

Corroboration: Confirmation or support of a witness's statement or other fact.

Costs: An allowance for expenses in prosecuting or defending a suit. Ordinarily, this does not include attorneys' fees.

Counsel: A legal representative, attorney, lawyer.

Counsel table: The physical location where the defense and prosecuting parties are seated throughout the duration of the trial.

Count: Each of the allegations of an offense listed in a charging document.

Counterclaim: A claim presented by a defendant in a civil lawsuit against the plaintiff. In essence, a counter lawsuit within a lawsuit.

Counterfeit: To forge, copy, or imitate, without authority or right and with the purpose to deceive or defraud by passing off the copy as genuine.

Court: 1. A unit of the judiciary authorized to decide disputed matters of fact, cases, or controversies. 2. Figuratively, the judge or judicial officer. Judges sometimes use *court* to refer to themselves in the third person, as in *the court has read the briefs*.

Court administrator/clerk of court: An officer appointed by the court or elected to oversee the administrative, nonjudicial activities of the court.

Court, appeals: In some states, the highest appellate court, where it is the court's discretion whether to hear the case on appeal.

Court-appointed counsel: A defense attorney designated by the court to represent a defendant who does not have the funds to retain an attorney.

Court costs: The expenses of prosecuting or defending a lawsuit, other than the attorneys' fees. An amount of money may be awarded to the successful party (and may be recoverable from the losing party) as reimbursement for court costs.

Court, district: 1. Federal: a trial court with general federal jurisdiction. 2. State: meaning varies from state to state.

Court, municipal: A court having jurisdiction (usually civil and criminal) over cases arising within the city or community in which it sits.

Court of record: A court in which the proceedings are recorded, transcribed, and maintained as permanent records.

Court order: A written direction or command delivered by a court or judge.

Court reporter: A person who makes a word-for-word record of what is said in court and produces a transcript of the proceedings upon request.

Courtroom: The section of a courthouse in which the judge presides over the proceedings.

Credibility: The quality in a witness that makes his or her testimony believable.

Criminal case: A case brought by the government against a person accused of committing a crime.

Cross-claim: A claim by a codefendant or coplaintiffs against each other and not against persons on the opposite side of the lawsuit.

Cross-examination: The questioning of a witness produced by the other side.

Damages: Money awarded by a court to a person injured by the unlawful act or negligence of another person.

***Daubert* motion:** A motion, raised before or during trial, to exclude the presentation of unqualified evidence to the jury. This is a special case of motion *in limine*, usually used to exclude the testimony of an expert witness who has no such expertise or used questionable methods to obtain the information.

Decision: The judgment reached or given by a court of law.

Declaratory judgment: A judgment of the court that explains what the existing law is or expresses the opinion of the court without the need for enforcement.

Decree: An order of the court. A final decree is one that fully and finally disposes of the litigation. An interlocutory decree is a preliminary order that often disposes of only part of a lawsuit.

Defamation: That which tends to injure a person's reputation. *Libel* is published defamation and *slander* is spoken defamation.

Default: A failure to respond to a lawsuit within the specified time.

Default judgment: A judgment entered against a party who fails to appear in court, respond to the charges, or does not comply with an order, especially an order to provide or permit discovery.

Defendant: 1. In a criminal case, the person accused of the crime. 2. In a civil case, the person being sued.

Defense: 1. Defendant's statement of a reason why the plaintiff or prosecutor has no valid case against the defendant, especially a defendant's answer, denial, or plea. 2. Defendant's method and strategy in opposing the plaintiff or the prosecution. 3. One or more defendants in a trial.

Defense attorney: An attorney who represents the defendant.

Deliberate: 1. To discuss, ponder, or reflect upon before reaching a decision. A judge will usually deliberate before announcing a judgment. 2. Intentional, characterized by consideration and awareness.

Deliberation: The jury's decision-making process after hearing the evidence and closing arguments and being given the court's instructions.

De novo: New. A trial de novo is a new trial of a case.

Deposition: A pretrial discovery device by which one party questions the other party or a witness for the other party. It usually takes place in the office of one of the lawyers, in the presence of a court reporter, who transcribes what is said. Questions are asked and answered orally as if in court, with opportunity given to the adversary to cross-examine. Occasionally, the questions are submitted in writing and answered orally.

Direct evidence: Proof of facts by witnesses who saw acts done or heard words spoken.

Direct examination: The first questioning of witnesses by the party on whose behalf they are called.

Directed verdict: Now called judgment as a matter of law. An instruction by the judge to the jury to return a specific verdict.

Disbarment: Form of discipline of a lawyer resulting in the loss (often permanently) of that lawyer's right to practice law. It differs from censure (an official reprimand or condemnation) and from suspension (a temporary loss of the right to practice law).

Discovery: The procedure by which one or both parties disclose evidence that will be used at trial. The specific tools of discovery include depositions, interrogatories, and motions for the production of documents.

Dismiss: To terminate legal action involving outstanding charges against a defendant in a criminal case.

Dismissal with prejudice: The dismissal of a case, by which the same cause of action cannot be brought against the defendant again at a later date.

Dismissal without prejudice: The dismissal of a case without preventing the plaintiff from bringing the same cause of action against the defendant in the future.

Disposition: A final settlement or determination. The court decision terminating proceedings in a case before judgment is reached, or the final judgment.

Dissent: 1. To disagree. 2. An appellate court opinion setting forth the minority view and outlining the disagreement of one or more judges with the decision of the majority.

Dissolution: The act of bringing to an end; termination—for example, the dissolution of a marriage or other relationship.

District attorney: A lawyer appointed or elected to represent the state in criminal cases in his or her respective judicial district. See *prosecutor.*

Docket: A list of cases to be heard by a court, or a log containing brief entries of court proceedings.

Docket number: The designation assigned to each case filed in a particular court. Also called a case number.

Driving while intoxicated (DWI): The unlawful operation of a motor vehicle while under the influence of drugs or alcohol. In some jurisdictions it is synonymous with driving under the influence (DUI), but, in others, driving while intoxicated is a more serious offense than driving under the influence.

Drunk driving: The operation of a vehicle in an impaired state after consuming an amount of alcohol that, when tested, is above the state's legal alcohol limit.

Due process of law: The right of all persons to receive the guarantees and safeguards of the law and the judicial process. It includes such constitutional requirements as adequate notice, assistance of counsel, the right to remain silent, the right to a speedy and public trial, the right to an impartial jury, and the right to confront and secure witnesses.

Estoppel: A person's own act or acceptance of facts that precludes his or her later making claims to the contrary.

Et al.: An abbreviation for *et alia* (and others).

Et seq.: An abbreviation for *et sequentes* or *et sequential* (and the following); ordinarily used in referring to a section of statutes.

Evidence: Information presented in testimony or in documents that is used to persuade the fact finder (judge or jury) to decide the case for one side or the other.

Examination, direct: The first examination of a witness by the counsel who called the witness to testify.

Examination, recross: A second examination of a witness by the opposing counsel after the second examination (or redirect examination) by the counsel who called the witness to testify is completed.

Examination, redirect: A second examination of a witness by the counsel who called the witness to testify. This examination is usually focused on certain matters that were discussed by the opposing counsel's examination.

Exceptions: Declarations by either side in a civil or criminal case reserving the right to appeal a judge's ruling upon a motion. Also, in regulatory cases, objections by either side to points made by the other side or to rulings by the agency or one of its hearing officers.

Exclusion of witnesses: An order of the court requiring all witnesses to remain outside the courtroom until each is called to testify, except the plaintiff or defendant. The witnesses are ordered not to discuss their testimony with each other and may be held in contempt if they violate the order.

Exclusionary rule: The rule preventing illegally obtained evidence to be used in any trial.

Exclusive jurisdiction: The matter can only be filed in one court.

Ex contractu: Arising from a contract.

Exculpatory evidence: Evidence that tends to indicate that a defendant did not commit the alleged crime.

Ex delicto: Arising from a wrong; breach of duty. See *tort.*

Exhibit: A document or other item introduced as evidence during a trial or hearing.

Exonerate: To remove of a charge, responsibility, or duty.

Ex parte: On behalf of only one party, without notice to any other party. For example, a request for a search warrant is an ex parte proceeding because the person subject to the search is not notified of the proceeding and is not present at the hearing.

Ex parte proceeding: The legal procedure in which only one side is represented. It differs from an adversary system or adversary proceeding.

Expert: See *expert witness.*

Expert testimony: Testimony given in relation to some scientific, technical, or professional matter by an expert (i.e., person qualified to speak authoritatively by reason of his or her special training, skill, or familiarity with the subject).

Expert witness: A person who has a defined level of expertise in his or her chosen field of science, technology, business, religion, or the arts that requires an opinion.

Ex post facto: After the fact. The Constitution prohibits the enactment of ex post facto laws. These are laws that permit conviction and punishment for a lawful act performed before the law was changed and the act made illegal.

Expungement: Official and formal erasure of a record or partial contents of a record.

Fee simple: The most complete, unlimited form of ownership of real property, which endures until the current holder dies without heir.

Fiduciary: A person having a legal relationship of trust and confidence to another and having a duty to act primarily for the other's benefit (e.g., a guardian, trustee, or executor).

File: To place a paper in the official custody of the clerk of court to enter into the files or records of a case.

Finding: Formal conclusion by a judge or jury on issues of fact.

Foundation: In a trial, a foundation must be laid to establish the basis for the admissibility of certain types of evidence. For example, the qualifications of an expert witness must be presented to the court and found acceptable before expert testimony will be admissible.

General assignment: The voluntary transfer, by a debtor, of all property to a trustee for the benefit of all of his or her creditors.

General jurisdiction: Refers to courts that have no limit on the types of criminal and civil cases they may hear.

Grounds: A foundation or basis; points relied on.

Guilty: Responsible for a delinquency, crime, or other offense; not innocent.

Habeas corpus: A writ commanding that a party be brought before a court or judge to protect him or her from unlawful imprisonment or custody.

Hearing, contested: A hearing held for the purpose of deciding issues or facts of law that both parties are disputing.

Hearing de novo: A full new hearing.

Hearing, preliminary: The hearing given to a person accused of a crime, by a magistrate or judge, to determine whether there is enough evidence to warrant the confinement of and holding to bail the person accused.

Hearsay evidence: Recounting of events by a witness who did not see or hear the incident in question but heard about it from someone else; not usually admissible in court.

Hostile witness: A witness whose testimony is not favorable to the party who calls him or her as a witness. A hostile witness may be asked leading questions and may be cross-examined by the party who calls him or her to the stand. An *expert witness* may be declared a hostile witness.

Hung jury: A jury whose members cannot agree upon a verdict.

Hypothetical question: An imaginary situation, incorporating facts previously admitted into evidence, upon which an *expert witness* is permitted to give an opinion as to a condition resulting from the situation.

Immunity: Grant by the court that ensures that someone will not face prosecution in return for providing evidence in a criminal proceeding.

Impanel: To seat a jury. When *voir dire* is finished and both sides have exercised their challenges, the jury members are sworn in and the trial is ready to proceed.

Impeachment of witness: To call into question the truthfulness of a witness.

Implied contract: A contract in which the promise made by the obligor is not expressed, but inferred by one's conduct or implied in law.

Inadmissible: That which, under the rules of evidence, cannot be admitted as evidence in a trial or hearing.

In camera: In chambers or in private. A hearing in camera takes place in the judge's office outside the presence of the jury and the public.

Incapacity: The lack of power or the legal ability to act.

Indictment: A formal written accusation, issued by a grand jury, charging a party with a crime.

In forma pauperis: Permission given to a person who cannot afford to pay to proceed in a lawsuit without having to pay court fees. It ensures that one is not deprived of the rights to litigate and appeal because of financial standing.

Information: A formal written document filed by the prosecutor detailing the criminal charges against the defendant. An alternative to an indictment, it serves to bring a defendant to trial.

Injunction: Writ or order by a court prohibiting a specific action from being carried out by a person or group.

In limine: (Latin: at the threshold): *motion in limine* is a motion, made before the start of a trial, requesting that the judge rule that certain evidence may, or may not, be introduced to the jury in a trial; this is done in the judge's chambers, out of hearing of the jury, to shield the jury from possibly inadmissible and harmful evidence.

Innocent until proven guilty: A belief in the American legal system that states that all people accused of a criminal act are considered not to have committed the crime until the evidence leaves no doubt in the mind of the court or the jury that the accused did or did not commit the crime.

In propia persona: In courts, this refers to a person who presents his or her own case without a lawyer. See *pro per* and *pro se.*

In rem: A procedural term used to designate proceedings or actions instituted against the thing, in contrast to actions instituted in personam, or against the person.

Instructions: Judge's explanation to the jury before it begins deliberations of the questions it must answer and the applicable law governing the case. Also called *charge.*

Intangible assets: Nonphysical items such as stock certificates, bonds, bank accounts, and pension benefits that have value and must be taken into account in estate planning.

Intent: The purpose to use a particular means to bring about a certain result.

Inter alia: Among other things.

Interlocutory: Provisional; not final. An interlocutory order or an interlocutory appeal concerns only a part of the issues raised in a lawsuit. Compare to *decree.*

Interrogatories: Written questions asked by one party in a lawsuit for which the opposing party must provide written answers.

Intervention: An action by which a third person who may be affected by a lawsuit is permitted to become a party to the suit; differs from the process of becoming an *amicus curiae.*

Inter vivos gift: A gift made during the giver's life.

Inter vivos trust: Another name for a living trust.

Investigation: A legal inquiry to discover and collect facts concerning a certain matter.

Irrelevant: Evidence not sufficiently related to the matter in issue.

Issue: 1. The disputed point in a disagreement between parties in a lawsuit. 2. To send out officially, as in *to issue an order.*

Joint venture: An association of persons jointly undertaking some commercial enterprise. Unlike a partnership, a joint venture does not entail a continuing relationship among the parties.

Judge: An elected or appointed public official with authority to hear and decide cases in a court of law.

Judgment: The final decision of the court, resolving the dispute; an opinion; an award.

Judicial notice: A court's recognition of the truth of basic facts without formal evidence.

Judicial review: The authority of a court to review the official actions of other branches of government. Also, the authority to declare unconstitutional the actions of other branches.

Jurisdiction: 1. The legal authority of a court to hear and decide a case. 2. The geographic area over which the court has authority to decide cases.

Jurisprudence: The study of law and the structure of the legal system.

Juror: Member of a jury.

Juror, alternate: Additional juror impaneled in case of sickness or disability of another juror.

Jury: A body of persons temporarily selected from the citizens of a particular district sworn to listen to the evidence in a trial and declare a verdict on matters of fact.

Jury box: The specific place in the courtroom where the jury sits during the trial.

Jury commissioner: The court officer responsible for choosing the panel of persons to serve as potential jurors for a particular court term.

Jury foreman: The juror who chairs the jury during deliberations and speaks for the jury in court when announcing the verdict.

Jury, hung: A jury that is unable to agree on a verdict after a suitable period of deliberation.

Jury trial: Trial in which a jury decides issues of fact, as opposed to trial only before a judge.

Justiciable: Issues and claims capable of being properly examined in court.

Law: The combination of those rules and principles of conduct promulgated by legislative authority, derived from court decisions, and established by local custom.

Law and motion: A setting before a judge at which time a variety of motions, pleas, sentencing, and orders to show cause or procedural requests may be presented. Normally, evidence is not taken. Defendants must be present.

Law clerks: Persons trained in the law who assist judges in researching legal opinions.

Lawsuit: An action between two or more persons in the courts of law, not a criminal matter.

Lay person: One not trained in law.

Leading question: A question that instructs the witness how to answer or puts words in his mouth to be echoed back. A question that suggests the desired answer to the witness.

Libel: Published words or pictures that falsely and maliciously harm the reputation of a person. See *defamation.*

Lien: A legal claim against another person's property as security for a debt. A lien does not convey ownership of the property but gives the lien holder a right to have his or her debt satisfied out of the proceeds of the property if the debt is not otherwise paid.

Limine: A motion requesting that the court not allow certain evidence that might prejudice the jury.

Limited action: A civil action in which recovery of less than a certain amount (as specified by statute) is sought. Simplified rules of procedure are used in such actions.

Limited jurisdiction: Refers to courts that are limited in the types of criminal and civil cases they may hear. For example, traffic violations generally are heard by limited jurisdiction courts.

Lis pendens: A pending suit.

Litigant: A party to a lawsuit. Litigation refers to a case, controversy, or lawsuit.

Litigation: A lawsuit.

Magistrate: Judicial officer exercising some of the functions of a judge. It also refers in a general way to a judge.

Malfeasance: Evil doing; ill conduct; the commission of some act that is positively prohibited by law.

Malice: Ill will, hatred, or hostility by one person toward another, which may prompt the intentional doing of a wrongful act without legal justification or excuse.

Malicious mischief: Willful destruction of property from actual ill will toward or resentment of its owner or possessor.

Malicious prosecution: An action instituted with intention of injuring the defendant and without probable cause that terminates in favor of the person prosecuted.

Malpractice: Violation of a professional duty to act with reasonable care and in good faith without fraud or collusion. This term is usually applied to such conduct by doctors, lawyers, or accountants.

Mandamus: A writ issued by a court ordering a public official to perform an act.

Mandate: A judicial command or order proceeding from a court or judicial officer directing the proper officer to enforce a judgment, sentence, or decree.

Material evidence: That quality of evidence that tends to influence the trier of fact because of its logical connection with the issue.

Mediation: A form of alternative dispute resolution in which the parties bring their dispute to a neutral third party, who helps them agree on a settlement.

Memorialize: To mark by observation in writing.

Merits: Strict legal rights of the parties; a decision on the merits is one that reaches the right(s) of a party, as distinguished from disposition of a case on a ground not reaching the right(s) raised in an action; for example, entry of nolle prosequi before a criminal trial begins is a disposition other than on the merits, allowing trial on those charges at a later time without double jeopardy attaching. Similarly, dismissal of a civil action on a preliminary motion raising a technicality, such as improper service of process, does not result in res judicata of an issue.

Mistrial: An invalid trial caused by some legal error. When a judge declares a mistrial, the trial must start again from the beginning, including the selection of a new jury.

Mitigating circumstances: Those circumstances that do not constitute a justification or excuse for an offense but may be considered as reasons for reducing the degree of blame.

Mitigating factors: Facts that do not constitute a justification or excuse for an offense but that may be considered as reasons for reducing the degree of blame.

Modification: A change, alteration, or amendment that introduces new elements into the details, or cancels some of them, but leaves the general purpose and effect of the subject matter intact.

Moot: A moot case or a moot point is one not subject to a judicial determination because it involves an abstract question or a pretended controversy that has not yet actually arisen or has already passed. Mootness usually refers to a court's refusal to consider a case because the issue involved has been resolved prior to the court's decision, leaving nothing that would be affected by the court's decision.

Moral turpitude: Immorality. An element of crimes inherently bad, as opposed to crimes bad merely because they are forbidden by statute.

Motion: Oral or written request made by a party to an action before, during, or after a trial asking the judge to issue a ruling or order in that party's favor.

Motion denied: Ruling or order issued by the judge denying the party's request.

Motion granted: Ruling or order issued by the judge granting the party's request.

Motion in limine: A written motion, usually made before or after the beginning of a jury trial, for a protective order against prejudicial questions and statements.

Multiplicity of actions: Numerous and unnecessary attempts to litigate the same issue.

Ne exeat: A writ that forbids the person to whom it is addressed to leave the country, the state, or the jurisdiction of the court.

Negligence: Failure to exercise the degree of care that a reasonable person would use under the same circumstances.

Next friend: One acting without formal appointment as guardian for the benefit of an infant, a person of unsound mind not judicially declared incompetent, or other person under some disability.

No bill: This phrase, endorsed by a grand jury on the written indictment submitted to it for its approval, means that the evidence was found insufficient to indict.

No-contest clause: Language in a will that provides that a person who makes a legal challenge to the will's validity will be disinherited.

No-fault proceedings: A civil case in which parties may resolve their dispute without a formal finding of error or fault.

Nolle prosequi: Translated: "I do not choose to prosecute." A decision by a prosecutor not to go forward with charging a crime. Also loosely called *nolle pros.*

Nolo contendere: A plea of no contest. In many jurisdictions, it is an expression that the matter will not be contested but without an admission of guilt. In other jurisdictions, it is an admission of the charges and is equivalent to a guilty plea.

Nominal party: One who is joined as a party or defendant merely because the technical rules of pleading require his presence in the record.

Non compos mentis: Not of sound mind; insane.

Non est inventus: Translated: "not to be found." A process when service is not made because the person to be served was not found.

Non obstante verdicto (NOV): "Notwithstanding the verdict." A verdict entered by the judge contrary to a jury's verdict.

Not guilty: The form of verdict in criminal cases where the jury acquits the defendant and finds him or her not guilty.

Notice: Formal notification to the party who has been sued in a civil case of the fact that the lawsuit has been filed. Also, any form of notification of a legal proceeding.

Notice to produce: In practice, a notice in writing requiring the opposite party to produce a certain described paper or document at the trial or in the course of pretrial discovery.

Null and void: Having no force, legal power to bind, or validity.

Nunc pro tunica: Legal phrase applied to acts that are allowed after the time when they should be done, with a retroactive effect.

Oath: Written or oral pledge by a witness to speak the truth.

Object: To protest to the court against an act or omission by the opposing party.

Objection: A protest to the court against an act or omission by the opposing party.

Objection overruled: A ruling by the court upholding the act or omission of the opposing party.

Objection sustained: A ruling by the court in favor of the party making the objection.

Of counsel: A phrase commonly applied to counsel employed to assist in the preparation or management of the case or its presentation on appeal, but who is not the principal attorney for the party.

Offer of proof: Presentation of evidence to the court (out of the hearing of the jury) for the court's decision of whether the evidence is admissible.

Opening argument: The initial statement made by attorneys for each side, outlining the facts each intends to establish during the trial.

Opening statement: See *opening argument.*

Opinion: A judge's written explanation of a decision of the court or of a majority of judges. A dissenting opinion disagrees with the majority opinion because of the reasoning or the principles of law on which the decision is based. A concurring opinion agrees with the decision of the court but offers further comment. A *per curiam* opinion is an unsigned opinion *of the court* or *through the court.*

Opinion evidence: Witnesses are normally required to confine their testimony to statements of fact and are not allowed to give their opinions in court. However, if a witness is qualified as an expert in a particular field, he or she may be allowed to state an opinion as an expert based on certain facts.

Opposing counsel: In the current context, the attorney who represents the *other side* of the case.

Oral argument: An opportunity for lawyers to summarize their position before the court and also to answer the judge's questions.

Order, court: A written or verbal command from a court directing or forbidding an action.

Order to show cause: Court order requiring appearance to show cause why the court should not take a particular course of action. If the party fails to appear or to give sufficient reasons why the court should take no action, the court will take the action. In criminal cases, the defendant must show why probation should not be revoked.

Ordinance: An act of legislation of a local governing body such as a city, town, or county.

Original jurisdiction: The court in which a matter must first be filed.

Overrule: A judge's decision not to allow an objection. A decision by a higher court finding that a lower court decision was wrong.

Overruled: See *overrule.*

Overt act: An open act showing the intent to commit a crime.

Paralegal (aka paralegal assistant): A person with legal skills who is not an attorney and works under the supervision of a lawyer or is otherwise authorized by law to use those legal skills.

Party: A person, business, or government agency actively involved in the prosecution or defense of a legal proceeding.

Patent: A government grant giving an inventor the exclusive right to make or sell his or her invention for a term of years.

Pending: Begun, but not yet completed. Thus, an action is pending from its inception until the rendition of its final judgment.

Per curium opinion: An unsigned opinion of the court.

Peremptory challenge: The right to challenge a juror without assigning a reason for the challenge.

Perjury: The act of lying or making verifiably false statements under oath or affirmation in a court of law or in any of various sworn statements in writing; the rules for perjury also apply to witnesses who have affirmed that they are telling the truth.

Permanent injunction: A court order requiring that some action be taken or that some party refrain from taking action. It differs from forms of temporary relief, such as a temporary restraining order or preliminary injunction.

Personal property: Tangible physical property (such as cars, clothing, furniture, and jewelry) and intangible personal property. This does not include real property such as land or rights in land.

Personal representative: The person who administers an estate. If named in a will, that person's title is "executor." If there is no valid will, that person's title is "administrator."

Petition: A formal, written application to the court requesting judicial action on some matter.

Petitioner: The person filing an action in a court of original jurisdiction. Also, the person who appeals the judgment of a lower court. The opposing party is called the respondent.

Petit jury: The ordinary jury of twelve (or fewer) persons for the trial of a civil or criminal case. So called to distinguish it from the grand jury.

Plaintiff: A person who initiates a lawsuit against another. Also called the complainant.

Plea: In a criminal proceeding, the defendant's declaration in open court that he or she is guilty or not guilty. The defendant's answer to the charges made in the indictment or information.

Plea bargain: The process whereby the accused and the prosecutor in a criminal case work out a mutually satisfactory disposition of the case subject to court approval. Usually involves the defendant's pleading guilty to a lesser offense or to only one.

Pleadings: The written statements of fact and law filed by the parties to a lawsuit.

Polling the jury: The act, after a jury verdict has been announced, of asking jurors individually whether they agree with the verdict.

Post conviction: A procedure by which a convicted defendant challenges the conviction or sentence on the basis of some alleged violation or error.

Postponement: To put off or delay a court hearing.

Power of attorney: Formal authorization of a person to act in the interest of another person.

Precedent: A previously decided case that guides the decision of future cases.

Pre-injunction: Court order requiring action or forbidding action until a decision can be made whether to issue a permanent injunction. It differs from a temporary restraining order.

Prejudice: A forejudgment, bias, or preconceived opinion.

Prejudicial error: Synonymous with reversible error; an error that warrants the appellate court in reversing the judgment before it.

Prejudicial evidence: Evidence that might unfairly sway the judge or jury to one side or the other.

Preliminary examination: The hearing available to a person charged with a felony to determine if there is enough evidence (probable cause) to hold him or her for trial.

Preliminary hearing: Another term for arraignment.

Preliminary injunction: In civil cases when it is necessary to preserve the status quo prior to trial, the court may issue a preliminary injunction or temporary restraining order ordering a party to carry out a specified activity.

Preponderance of evidence: Evidence that is of greater weight or more convincing than the evidence offered in opposition to it.

Presumption: An inference of the truth or falsity of a proposition or fact that stands until rebutted by evidence to the contrary.

Presumption of innocence: A hallowed principle of criminal law that a person is innocent of a crime until proven guilty. The government has the burden of proving every element of a crime beyond a reasonable doubt, and the defendant has no burden to prove his innocence.

Presumption of law: A rule of law that courts and judges must draw a particular inference from a particular fact or from particular evidence.

Pretrial conference: A meeting between the judge and the lawyers involved in a lawsuit to narrow the issues in the suit, agree on what will be presented at the trial, and make a final effort to settle the case without a trial.

Prima facie case: A case that is sufficient and has the minimum amount of evidence necessary to allow it to continue in the judicial process.

Privilege: A legal right, exemption, or immunity granted to a person, company, or class that is beyond the common advantages of other citizens.

Privileged communication: Confidential communication to a person or persons protected by law against any disclosure, including forced disclosure in legal proceedings; communications between lawyer and client are typically privileged.

Privileged information: Information that a person or an organization has a right to withhold from judicial proceedings.

Probable cause: A reasonable belief that a crime has been or is being committed; the basis for all lawful searches, seizures, and arrests.

Pro bono publico: "For the public good." Lawyers representing clients without a fee are said to be working *pro bono publico*; often shortened to *pro bono.*

Procedural law: The method established normally by rules to be followed in a case; the formal steps in a judicial proceeding.

Proffer: An offer of proof as to what the evidence would be if a witness were called to testify or answer a question.

Proof: Any fact or evidence that leads to a judgment of the court.

Pro per: One who represents oneself in a court proceeding without the assistance of a lawyer. Also known as *pro se*. See also *in propia persona.*

Pro se: A Latin term meaning "on one's own behalf"; in courts, it refers to persons who present their own cases without lawyers. See *in propia persona* and *pro per.*

Prosecuting attorney: See *prosecutor* and *district attorney.*

Prosecution: A proceeding instituted and carried on in order to determine the guilt or innocence of the accused.

Prosecutor: A trial lawyer representing the government in a criminal case and the interests of the state in civil matters. In criminal cases, the prosecutor has the responsibility of deciding whom and when to prosecute.

Protective order: A court order to protect a person from further harassment, service of process, or discovery.

Proximate cause: The act that caused an event to occur. A person generally is liable only if an injury was proximately caused by his or her action or by his or her failure to act when he or she had a duty to act.

Public defender: An attorney appointed by a court or employed by a government agency whose work consists primarily of defending people who are unable to hire a lawyer due to economic reasons.

Punitive damages: Money awarded to an injured person, over and above the measurable value of the injury, in order to punish the person who hurt him.

Purge: To clean or clear, such as eliminating inactive records from court files; with respect to civil contempt, to cure the noncompliance that caused the contempt finding.

Quantum meruit: Expression meaning "as much as he deserves" that describes the extent of liability on a contract implied by law.

Quash: To overthrow, vacate, annul, or make void.

Quasi-judicial: Authority or discretion vested in an officer whose acts partake of a judicial character.

Quid pro quo: "What for what"; something for something; giving one valuable thing for another.

Quo warranto: Writ issuable by the state through which it demands an individual to show by what right he or she exercises an authority that can only be exercised through grant or franchise from the state or why he or she should not be removed from office.

Ratification: The confirmation or adoption of a previous act done either by the party or by another.

Ratio decidendi: The ground or reason of the decision in a case.

Real evidence: Evidence given to explain, repel, counteract, or disprove facts given in evidence by the adverse party.

Reasonable doubt, beyond a: The degree of certainty required for a juror to find a criminal defendant legally guilty. An accused person is entitled to acquittal if, in the minds of the jury, his or her guilt has not been proved beyond a *reasonable doubt*; that state of mind of jurors in which they cannot say they feel a persisting conviction as to the truth of the charge.

Reasonable person: A phrase used to denote a hypothetical person who exercises qualities of attention, knowledge, intelligence, and judgment that society requires of its members for the protection of his or her own interest and the interests of others. Thus, the test of negligence is based on either a failure to do something that a reasonable person, guided by considerations that ordinarily regulate conduct, would do, or on the doing of something that a reasonable and prudent (wise) person would not do.

Rebuttal: Evidence given to explain, counteract, or disprove facts given by the opposing counsel.

Recall: Cancellation by a court of a warrant before its execution by the arrest of a defendant; also, a process by which a retired judge may be asked to sit on a particular case.

Record: All the documents and evidence plus transcripts of oral proceedings in a case.

Recuse: The process by which a judge is disqualified from hearing a case, on his or her own motion or upon the objection of either party.

Redirect examination: Opportunity to present rebuttal evidence after one's evidence has been subjected to cross-examination.

Redress: To set right; to remedy; to compensate; to remove the causes of a grievance.

Referee: A person to whom the court refers a pending case to take testimony, hear the parties, and report back to the court. A referee is an officer with judicial powers who serves as an arm of the court.

Regulation: A rule or order prescribed for management or government.

Rehearing: Another hearing of a civil or criminal case by the same court in which the case was originally heard.

Rejoinder: Opportunity for the side that opened the case to offer limited response to evidence presented during the rebuttal by the opposing side.

Relevant evidence: Evidence that helps to prove a point or issue in a case.

Remittitur: The reduction by a judge of the damages awarded by a jury.

Removal: The transfer of a state case to federal court for trial—in civil cases because the parties are from different states; in criminal and some civil cases because there is a significant possibility that there could not be a fair trial in state court.

Replevin: An action for the recovery of a possession that has been wrongfully taken.

Reply: The response by a party to charges raised in a pleading by the other party.

Report: An official or formal statement of facts or proceedings.

Res ipsa loquitur: Literally, "a thing that speaks for itself." In tort law, the doctrine that holds a defendant guilty of negligence without an actual showing that he or she was negligent.

Res judicata: A rule of civil law that once a matter has been litigated and final judgment has been rendered by the trial court, the matter cannot be relitigated by the parties in the same court or any other trial court.

Respondeat superior: "Let the master answer." This doctrine holds that employers are responsible for the acts and omissions of their employees and agents when they are done within the scope of the employees' duties.

Respondent: The party who makes an answer to a bill or other proceedings in equity; also refers to the party against whom an appeal is brought. Sometimes called an appellee.

Rest: A party is said to rest or rest its case when it has presented all the evidence it intends to offer.

Restitution: Act of giving the equivalent for any loss, damage, or injury.

Retainer: Act of the client in employing the attorney or counsel. Also denotes the fee the client pays when he or she retains the attorney to act for him or her.

Reverse: An action of a higher court in setting aside or revoking a lower court decision.

Reversible error: A procedural error during a trial or hearing sufficiently harmful to justify reversing the judgment of a lower court. See *prejudicial error.*

Rule: An established standard, guide, or regulation.

Rule of court: An order made by a court having competent jurisdiction. Rules of court are either general or special; the former are the regulations by which the practice of the court is governed, and the latter are special orders made in particular cases.

Rules of evidence: Standards governing whether evidence in a civil or criminal case is admissible.

Sanction: A punitive act designed to secure enforcement by imposing a penalty for its violation. For example, a sanction may be imposed for failure to comply with discovery orders.

Sealing: The closure of court records to inspection, except to the parties.

Secured debt: In bankruptcy proceedings, a debt is secured if the debtor gave the creditor a right to repossess the property or goods used as collateral.

Sentence: The judgment formally pronounced by the court or judge upon the defendant after his or her conviction by imposing a punishment to be inflicted in the form of either a fine, incarceration, or probation.

Sentence report: A document containing background material on a convicted person. It is prepared to guide the judge in the imposition of a sentence; sometimes called a presentence report.

Service: The delivery of a legal document, such as a complaint, summons, or subpoena, notifying a person of a lawsuit or other legal action taken against him or her. Service, which constitutes formal legal notice, must be made by an officially authorized person in accordance with the formal requirements of the applicable laws.

Service of process: Notifying a person that he or she has been named as a party to a lawsuit or has been accused of some offense. Process consists of a summons, citation, or warrant, to which a copy of the complaint is attached.

Settlement: An agreement between parties that dictates what is being received by one party from the other.

Settlor: The person who sets up a trust. Also called the grantor.

Show cause: An order requiring a person to appear in court and present reasons why a certain order, judgment, or decree should not be issued.

Sidebar: A conference between the judge and lawyers, usually in the courtroom out of earshot of the jury and spectators.

Sovereign immunity: The doctrine that the state or federal government is immune to lawsuit unless it gives its consent.

Standard of proof: There are essentially three standards of proof applicable in most court proceedings. In criminal cases, the offense must be proven beyond a reasonable doubt, the highest standard. In civil cases and neglect and dependency proceedings, the lowest standard applies by a mere preponderance of the evidence (more likely than not). In some civil cases and in juvenile proceedings such as a permanent termination of parental rights, an intermediate standard applies, proof by clear and convincing evidence.

Standing: The legal right to bring a lawsuit. Only a person with something at stake has standing to bring a lawsuit.

Stare decisis: The doctrine that courts will follow principles of law laid down in previous cases. Similar to precedent.

Statement, closing: The final statements by the attorneys to the jury or court summarizing the evidence that they have established and the evidence that the other side has failed to establish. Also known as closing argument.

Statement, opening: Outline or summary of the nature of the case and of the anticipated proof presented by the attorney to the jury before any evidence is submitted. Also known as opening argument.

Statute: A formal, written statement by a legislature declaring, commanding, or prohibiting something.

Statute of limitations: The time within which a plaintiff must begin a lawsuit (in civil cases) or a prosecutor must bring charges (in criminal cases). There are different statutes of limitations at both the federal and state levels for different kinds of lawsuits or crimes.

Statutory: Prescribed or authorized by statute or law, as in a statutory right.

Statutory law: Law enacted by the legislative branch of government, as distinguished from case law or common law.

Stay: The act of stopping a judicial proceeding by order of the court.

Stipulate: An agreement by attorneys on both sides of a civil or criminal case about some aspect of the case (e.g., to extend the time to answer, to adjourn the trial date, or to admit certain facts at the trial).

Strict liability: A concept applied by courts in product liability cases in which a seller is liable for any and all defective or hazardous products that unduly threaten a consumer's personal safety.

Sua sponte: A Latin phrase that means "on one's own behalf"; voluntary, without prompting or suggestion.

Sub curia: Translated: "under the law"; the holding of a case by a court under consideration, sometimes to await the filing of a document, such as a presentence investigation report or memorandum of law, or to write an opinion.

Subject matter jurisdiction: A determination of whether a court has jurisdiction, or the power to render decisions, over claims or disputes.

Subpoena: An order of the court that requires a person to be present at a certain time and place to give testimony upon a certain matter. Failure to appear may be punishable as contempt of court.

Subpoena duces tecum: Court order commanding a witness to bring certain documents or records to court.

Substantive law: The law dealing with rights, duties, and liabilities, as contrasted with procedural law, which governs the technical aspects of enforcing civil or criminal laws.

Sue: To commence legal proceedings for recovery of a right.

Suit: Any proceeding by one person or persons against another in a court of law.

Summary judgment: A decision made on the basis of statements and evidence presented for the record without a trial. It is used when there is no dispute as to the facts of the case, and one party is entitled to judgment as a matter of law.

Summons: A notice to a defendant that he or she has been sued or charged with a crime and is required to appear in court. A jury summons requires the person receiving it to report for possible jury duty.

Supersedeas: A writ issued by an appellate court to preserve the status quo pending review of a judgment or pending other exercise of its jurisdiction.

Suppress: To forbid the use of evidence at a trial because it is improper or was improperly obtained. See also *exclusionary rule.*

Suppression hearing: A hearing on a criminal defendant's motion to prohibit the prosecutor's use of evidence alleged to have been obtained in violation of the defendant's rights. This hearing is held outside the presence of the jury, either prior to or at trial. The judge must rule as a matter of law on the motion.

Sustain: To maintain, to affirm, to approve.

Swear: To put to oath and declare as truth.

Tangible: Capable of being perceived, especially by the sense of touch.

Tangible personal property memorandum (TPPM): A legal document referred to in a will and used to guide the distribution of tangible personal property.

Temporary relief: Any form of action by a court granting one of the parties an order to protect its interest pending further action by the court.

Temporary restraining order: A judge's order forbidding certain actions until a full hearing can be held; usually of short duration. Often referred to as a *TRO.*

Testify: To make a declaration under oath in a judicial inquiry for the purpose of establishing or proving some fact.

Testimony: The evidence given by a witness under oath. It does not include evidence from documents and other physical evidence.

Tort: A civil injury or wrong committed on the person or property of another. A tort is an infringement on the rights of an individual but not founded on a contract. The most common tort action is a suit for damages sustained in an automobile accident. See *ex delicto.*

Transcript: A written, word-for-word record of what was said, either in a proceeding such as a trial or during some other conversation, as in a transcript of a hearing or oral deposition.

Transitory: Actions are *transitory* when they might have taken place anywhere, and they are *local* when they could occur only in some particular place.

Trial: A judicial examination and determination of issues between parties before a court that has jurisdiction.

Trial court: See *trial, court (bench).*

Trial, court (bench): A trial where the jury is waived and the case is seen before the judge alone.

Trial de novo: A new trial or retrial held in an appellate court in which the whole case is heard as if no trial had been heard in the lower court or administrative agency.

Trial, speedy: The Sixth Amendment of the Constitution guarantees the accused an immediate trial in accordance with prevailing rules, regulations, and proceedings of law.

Trier of fact: This term includes the jury—or the judge in a jury-waived trial—who has the obligation to make findings of fact rather than rulings of law.

TRO: See temporary restraining order.

True test copy: A copy of a court document given under the clerk's seal but not certified.

Undue influence: Whatever destroys free will and causes a person to do something he would not do if left to himself.

Unjust enrichment, doctrine of: The principle that one person should not be permitted to unjustly enrich himself at the expense of another but should be required to make restitution for the property or benefit received.

Usury: Charging a higher interest rate or higher fees than the law allows.

Venire: A writ summoning persons to court to act as jurors; also refers to the people summoned for jury duty.

Venue: The proper geographic area (county, city, or district) in which a court with jurisdiction over the subject matter may hear a case.

Verdict: The opinion of a jury or a judge when there is no jury on the factual issues of a case.

Voir dire: *To speak the truth*; the preliminary examination that the court and attorneys make of prospective jurors to determine their qualification and suitability to serve as jurors.

Weight of the evidence: The persuasiveness of certain evidence when compared with other evidence that is presented.

With prejudice: Applied to orders of judgment dismissing a case, meaning that the plaintiff is forever barred from bringing a lawsuit on the same claim or cause.

Without prejudice: A claim or cause dismissed without prejudice may be the subject of a new lawsuit.

Witness: 1. One who testifies to what he or she has seen, heard, or otherwise observed. 2. To subscribe one's name to a document for the purpose of authenticity.

Witness, defense: A nonhostile witness called by the defense counsel to assist in proving the defense's case.

Witness, expert: A witness who is qualified by knowledge, skill, experience, training, or education to provide a scientific, technical, or specialized opinion of the subject about which he or she is to testify. That knowledge must generally be such as is not normally possessed by the average person.

Witness, hostile: A witness whose relationship to the opposing party is such that his or her testimony may be prejudiced against that party. A witness declared to be hostile may be asked leading questions and is subject to cross-examination by the party who called him or her.

Witness, material: A witness who can give testimony relating to a particular matter that very few others, if any, can give.

Witness, prosecution: The person whose complaint commences a criminal prosecution and whose testimony is mainly relied on to secure a conviction at the trial.

Witness stand: The space in the courtroom occupied by a witness while testifying.

Writ: A court's written order commanding the addressee to do or refrain from doing some specified act.

Writ of certiorari: A formal written order from a higher court to a lower court requesting a transcript of the proceedings of a previous case for review; most commonly used in the context of the Supreme Court agreeing to hear a case (granting cert) or refusing to take up a case (denying cert).

Writ of execution: A writ to put in force the judgment or decree of a court.

Bibliography and Additional Reading

The italicized words accompanying a legal reference indicate the subject matter that was decided in that particular case.

American Society for Testing and Materials. 2008. *Annual book of standards.* West Conshohocken, PA: American Society for Testing and Materials.

Balabanian, D. M. 1987. Medium v. tedium: Video depositions come of age. Practicing Law Institute/Litigation 328. Videotaped record of a deposition.

Bauman v. Centex Corp., 611 F.2d 1115 (5th Cir. 1980). *Facts relied on are of a type reasonably relied on by experts in the particular field.*

Beggs, G. J. *Novel Expert Evidence in Federal Civil Rights Litigation.* American University Law Review October (1995).

Berkshire Mutual Ins. Co. v. Moffett, 378 F.2d 7 2 d10 07 (5th Cir. 1967). *Testimony of witnesses.*

Bradley, M. D. 1983. *The scientist and engineer in court.* Washington, D.C.: American Geophysical Union.

Brent, R. 2006. The *Daubert* decision. *Pediatrics* 118:2222–2225.

Brewer, S. *Scientific Expert Testimony and Intellectual Due Process.* 107 Yale Law Journal 1535 (1998).

Brownstein, D. A. 1993. *Law for the expert witness.* Ann Arbor, MI: Lewis Publishers.

Broyles, K. E. *Taking the Courtroom into the Classroom: A Proposal for Educating the Lay Juror in Complex Litigation Cases.* 64 George Washington Law Review 714 (1996).

Bryan v. John Bean Division of FMC Corp., 566 F.2d 541 (5th Cir. 1978). *Expert's reliance on facts and data supplied by third parties.*

Cecil, J. S., and T. E. Wiliging. 1993. Court-appointed experts: Defining the role of experts appointed under Federal Rule of Evidence 706. In *Reference manual on scientific evidence,* ed. S. Cecil. Washington, D.C.: Federal Judicial Center.

Centola, G. D. 2004. Discovery of experts who are not expected to testify at trial. *Mealey's Emerging Toxic Torts* (February 6). http://www.rivkinradler.com/rivkinradler/ Publications/newformat/200402centola.shtml

Chicago & N. W. Ry. Co. v. Green, 174 F.2d 55 (8th Cir. 1947). *Testimony of witnesses.*

Clifford, R. C. 1991. *Qualifying and attacking expert witnesses.* Santa Ana, CA: James Publishing.

Collins, M. B. *Taking the Deposition (and Getting It Right).* 90 Illinois Bar Journal 323 (2002).

Cooper, J., E. A. Bennett, and H. L. Suckle. 1996. Complex scientific testimony: How do jurors make decisions? *Law and Human Behavior* 20:379.

Cowbell, S. E. *Pretrial Mediation of Complex Scientific Cases: A Proposal to Reduce Jury and Judicial Confusion.* 75 Chicago-Kent Law Review 981 (2000).

Cranor, C. 1993. *Regulating toxic substances.* New York: Oxford University Press.

______. 2005. Scientific inferences in the laboratory and the law. *American Journal of Public Health* 95:S121–S128.

Danois, D. 1995. *Computer Animation Helps Win Cases: By Visualizing Complex Evidence, Juror Comprehension Rises.* 18 Pennsylvania Law Weekly S8 (1995).
Daubert v. Merrell Dow Pharmaceuticals Co. 113 S.Ct. 2728, 2741–2743, 125 L.Ed.2d 469, 482–485 (1993).
Daubert v. Merrell Dow Pharmaceuticals, Inc., 43 F.3d 1311, 1316 (9th Cir. 1995).
Daubert v. Merrell Dow Pharmaceuticals, Inc., United States Supreme Court No. 92–102.
Dewitt, J. S. 1991. *Novel scientific evidence and the juror: A scientific psychological approach to the* Frye-*relevancy controversy.* PhD thesis, University of Nevada, Reno. University Microfilms, Ann Arbor, MI, 1993.
______. *Juries and Experts: Sensing and Comprehending in Cases That Hinge on Expert Testimony.* 3 Nevada Lawyer 18 (1995).
Doelle v. United States, 309 F.2d 396 (5th Cir. 1962). *The power of a trial judge to appoint an expert of his or her own choosing.*
Epstein, E. S. 2001. *The attorney–client privilege and the work-product doctrine,* 4th ed. Chicago: Section of Litigation, American Bar Association.
Feder, H. A. 1991. *Succeeding as an expert witness.* New York: Van Nostrand Reinhold.
Freedman, M. H. 1990. *Understanding lawyers' ethics.* New York: Bender Publishers.
Fugitt v. Jones, 549 F.2d 1001 (5th Cir. 1977). *The power of a trial judge to appoint an expert of his or her own choosing.*
Furst, A. 1995. The *Frye* rule is out: Is junk science in? *Journal of the American College of Toxicology* 14:61–68.
______. 1997. *The toxicologist as an expert witness.* Boca Raton, FL: CRC Press, Taylor & Francis Group.
Georgia-Pacific Corp. v. United States, 640 F.2d 328 (Ct. C1. 1980). *The power of a trial judge to appoint an expert of his or her own choosing.*
Gering, J. E. 2002. *Handbook of federal civil discovery and disclosure,* 2nd ed. St. Paul, MN: West Group.
Gillers, S. 1979. *The rights of lawyers and clients.* New York: Avon Books.
Gleeson, J. G. 1993. Science in the courtroom: Does *Daubert* warrant a change? *Mealey's Toxic Torts* 2(3).
Goodman, J. 1986. *Probabilistic scientific evidence: Jurors' inferences.* PhD thesis, University of Washington. University Microfilms International, Ann Arbor, MI, 1987.
Gray v. Shell Oil Co., 469 F.2d 742 (9th Cir. 1972), cert. denied, 412 U.S. 943 (1973). *Testimony of witnesses.*
Hack, S. 2005. Trial and error: The Supreme Court's philosophy of science. *American Journal of Public Health* 95:S66–S73.
Handberg, R. B. *Expert Testimony on Eyewitness Identification: A New Pair of Glasses for the Jury.* 32 American Criminal Law Review 1013 (1985).
Haney v. Mizell Memorial Hospital, 744 F.2d 1467 (11th Cis. 1984). *An expert is allowed to express an opinion even if it embraces the ultimate issue to be decided by the fact finder.*
Hartford Fire Ins. v. Cagle, 249 F.2d 241 (10th Cir. 1957). *Testimony of witnesses.*
Harvard Law Review. Developments in the Law: Confronting the New Challenges of Scientific Evidence. 108 Harvard Law Review 1481 (1995).
Haydock, R. S. 2002. *Discovery practice,* 4th ed. New York: Aspen Law & Business.
Huber, P. 1991. *Galileo's revenge: Junk science in the court room.* New York: Basic Books, Perseus Books Group.
Imwinkelried, E. J. *The Standard for Admitting Scientific Evidence: A Critique from the Perspective of Juror Psychology.* 28 Villanova Law Review 554 (1983).

______. *Judge Versus the Jury: Who Should Decide Questions of Preliminary Facts Conditioning the Admissibility of Scientific Evidence?* 25 William and Mary Law Review 577 (1984).

Jacobs, M. S. *Testing the Assumptions Underlying the Debate about Scientific Evidence: A Closer Look at Juror "Competence" and Scientific "Objectivity."* 25 Connecticut Law Review 1083 (1993).

Jasanoff, S. 1995. *Science at the bar. Law, science and technology in America.* Cambridge, MA: Harvard University Press.

______. 2005. Law's knowledge: Science for justice in legal settings. *American Journal of Public Health.* 95:S49–S58.

Johnson, M. T., C. Krafka, and J. S. A. Cecil. 2000. Expert testimony in federal civil trials: A preliminary analysis. Federal Judicial Center, Washington, D.C., 20002-8003. www.fjc.gov *Expert witnesses may deliver expert evidence about facts from the domain of their expertise.*

Juhnke, D. H. 2005. Discovery of databases in litigation. Computer Forensics Inc. http://www. forensics.com/pdf/Database_Discovery.pdf

Kaufman, H. H. 2001. The expert witness: Neither Frye nor Daubert solved the problem: What can be done? *Science and Justice* 41:7–20.

Klapmeier v. Telecheck International, Inc., 482 F.2d 247 (8th Cir. 1973). *Testimony of witnesses.*

Koppenhaver, K. M. 1990. How to be a credible witness. Available from Forensic Document Examiners, P.O. Box 324, Joppa, MD, 21085.

Lolie v. Ohio Brass CG., 502 F.2d 741 (7th Cir. 1974). *Testimony by experts.*

Louis, D. E. 1982. Guidelines for expert witnesses. *Journal of the Air Pollution Control Association* 32:1029–1030.

Maatschappij Voor Industriele Waarden N. V. v. A. O. Smith Corp., 590 F.2d 415 (26 Cir. 1978). *Testimony by experts.*

Mahle, S. 1999. The impact of *Daubert v. Merrell Dow Pharmaceuticals, Inc.* on expert testimony: With applications to securities litigation. *Florida Bar Journal* April.

Malone, D. M., and P. T. Hoffman. 2001. *The effective deposition: Techniques and strategies that work,* rev. 2d ed. Notre Dame, IN: National Institute for Trial Advocacy.

Martiniak, C. 2002. *How to take and defend depositions,* 3d ed. New York: Aspen Law & Business.

Matson, J. V. 1990. *Effective expert witnessing.* Chelsea, MI: Lewis Publishers.

McElhaney, J. W. 2003. Deposition goals: Develop a plan to get what you're after from witnesses in discovery. *ABA Journal* 89:30.

Montoya, J. *A Theory of Compulsory Process Clause Discovery Rights.* 70 Indiana Law Journal (1995).

Moran v. Ford Motor Co., 476 F.2d 289 (8th Cir. 1973). *Testimony by experts.*

Morris, W. A. 1990. The chemist as an expert witness. *Chemist* 7:21.

Nichols v. Marshall, 486 F.2d 791 (10th Cir. 1973). *Testimony of witnesses.*

Noona, J. M., and M. A. Knoerzer. 1989. The attorney–client privilege and corporate transactions: Counsel as keeper of corporate secrets. In *The Attorney–Client Privilege under Siege. Tort and Insurance Practice.* Lake Buena Vista, FL, May 10–14.

Pandia.com. A short and easy search engine tutorial. http://www.pandia.com/goalgetter/

Ponder v. Warren Tool Corp., 834 F.2d 1553 (10th Cir. 1987). *An expert is allowed to express an opinion even if it embraces the ultimate issue to be decided by the fact finder.*

Randolph v. Collectramatic, Inc., 590 F.2d 844 (10th Cir. 1979). *Testimony of witnesses.*

Raschke, H. H. 2007. Discovery ordered from consulting (nontestifying) expert. *Property Insurance Law Committee Newsletter,* American Bar Association Spring.

Rice, P. R. 1999. *Attorney–client privilege in the United States,* 2nd ed. Minneapolis, MN: West Group.

Rodriquez v. Olin Corp., 780 F.2d 491 (5th Cir. 1986). *Facts relied on are of a type reasonably relied on by experts in the particular field.*

Rose Hall, Ltd., v. Chase Manhattan Overseas Banking Corp., 576 F. Supp. 107 (D.C. Del. 1983), afffd, 740 F.2d 956-958 (3d Cir. cert. denied, 469 U.S.). *Facts relied on are of a type reasonably relied on by experts in the particular field.*

Saks, M. J. *Scientific Evidence and the Ethical Obligations of Attorneys.* 49 Cleveland State Law Review 421 (2001).

Sanders, J. *Scientifically Complex Cases, Trial by Jury, and the Erosion of Adversarial Processes.* 48 DePaul Law Review 355 (1998).

Schutz, J. S. *The Expert Witness and Jury Comprehension: An Expert's Perspective.* 7 Cornell Journal of Law and Public Policy 107 (1997).

Schutz, R. J., and M. R. Lueck. *Computer Animation Tutors Jury; In Complex Litigation, High-Tech Graphical Presentations Help the Jury Understand Difficult Issues.* 18 National Law Journal C1 (1995).

Scott v. Spanjer Bros., Inc., 298 F.2d 928 (2d Cir. 1962). *The power of a trial judge to appoint an expert of his or her own choosing.*

Sherman, J. D. 1983. Women as expert witnesses: Trials and tribulations. *Trial* August:47–48.

Shipp v. General Motors Corp., 750 F.2d 418 (5th Cir. 1985). *Testimony by experts.*

Shuman, D. W. 1996. Assessing the believability of expert witnesses: Science in the jury box. *Jurimetrics* 37:23.

Smith v. Hobart Manufacturing Co., 185 F. Supp. 751 (D.C. 1960). *Testimony by experts.*

Specht v. Jensen, 853 F.2d 805 (10th Cir. 1988). *An expert is allowed to express an opinion even if it embraces the ultimate issue to be decided by the fact finder.*

Speight, J. G. 2001. *Handbook of petroleum analysis.* New York: John Wiley & Sons, Inc.

______. 2002. *Handbook of petroleum product analysis.* Hoboken, NJ: John Wiley & Sons, Inc.

Thomas, W. A., ed. 1974. *Scientists in the legal system: Tolerable meddlers or essential contributors?* Ann Arbor, MI: Ann Arbor Science Press.

Tinkham, T., and W. J. Wernz. 1993. *Attorney–client privilege, confidentiality, and work product doctrine in Minnesota.* Minneapolis, MN: Dorsey & Whitney.

United States v. Barker, 553 F.2d 1013 (6th Cir. 1977). *Testimony by experts.*

United States v. Baskes, 649 F.2d 471 (7th Cir. 1980), cert. denied, 450 U.S. 1000 (1981). *Testimony of witnesses.*

United States v. Battle, 859 F.2d 56 (8th Cir. 1988). *An expert is allowed to express an opinion even if it embraces the ultimate issue to be decided by the fact finder.*

United States v. Jackson, 688 F.2d 1121 (7th Cir. 1982). *Testimony of witnesses.*

United States v. Rizzo, 492 F.2d 443 (95th Cir. 1974). *Testimony of witnesses.*

United States v. R. J. Reynolds Tobacco Co., 416 F. Supp. 313 (DCNY 1976). *Expert's reliance on facts and data supplied by third parties.*

United States v. Sowards, 370 F.2d 87 (10th Cir. 1966).

United States v. Wysocki, 4, 57 F.2d 1155 (5th Cir. 1972), cert. denied, 409 U.S. 859 (1972). *Testimony by experts.*

Virginia Electric & Power Co. v. Sun Shipbuilding & Dry Dock Co., 68 F.R.D. 397 (D.C. 1975). *Testimony by experts.*

Viterbo v. Dow Chemical Co., 646 F. Supp. 1420 (E.D. Tex. 1986), affd, 826 F.2d 420 (5th Cir. 1987). *Facts relied on are of a type reasonably relied on by experts in the particular field.*

Viterbo v. Dow Chemical Co., 826 F.2d 420 (5th Cir. 1987). *Facts relied on are of a type reasonably relied on by experts in the particular field.*

Walker, L., and J. Monahan. *Daubert and the Reference Manual: An Essay on the Future of Science in Law.* 82 Virginia Law Review 837–857 (1996).

Wang, C. C. K., and H. E. Parker. 1984. How to be an effective expert witness. *Chemical Engineering News* February 20:87–90.

Watts v. Cypress Hill, No. 06 C 3348, 2008 WL 697356 (N.D. Ill. Mar. 12, 2008). *An expert report can be denied if the report does not include the basis and reasons for the expert's opinion(s).*

White, J. S. 2000/2001. Selecting and retaining experts. Northern California Register of Experts and Consultants, the Bar Association of San Francisco, ix–xi.

Withers, K. J. Undated. Is digital different? Electronic disclosure and discovery in civil litigation. http://www.kenwithers.com/articles/bileta/elecdisc.htm

Zenith Radio Corp. v. Matsushita Elec. Indus. Co., Ltd., 505 F. Supp 1313 (D.C. Pa. 1980). *Facts relied on are of a type reasonably relied on by experts in the particular field.*

Zweifach, L. J., and G. Zweifach. 1994. Preparing to take and taking the deposition. Practicing Law Institute/Litigation 507.

Index

D

E

F

G

H

T

U

V

W